Selected Titles in This Series

27 **Shui Feng, Anna T. Lawniczak, and S. R. S. Varadhan, Editors,** Hydrodynamic limits and related topics, 2000

26 **Neal Madras, Editor,** Monte Carlo methods, 2000

25 **A. G. Ramm, P. N. Shivakumar, and A. V. Strauss, Editors,** Operator theory and its applications, 2000

24 **Edward Bierstone, Boris Khesin, Askold Khovanskii, and Jerrold E. Marsden, Editors,** The Arnoldfest: Proceedings of a conference in honour of V. I. Arnold for his sixtieth birthday, 1999

23 **Myrna H. Wooders, Editor,** Topics in mathematical economics and game theory. Essays in honor of Robert J. Aumann, 1999

22 **Graciela Chichilnisky, Editor,** Topology and markets, 1999

21 **Shigui Ruan, Gail S. K. Wolkowicz, and Jianhong Wu, Editors,** Differential equations with applications to biology, 1998

20 **Peter A. Fillmore and James A. Mingo, Editors,** Operator algebras and their applications II, 1998

19 **William G. Dwyer, Stephen Halperin, Richard Kane, Stanley O. Kochman, Mark E. Mahowald, and Paul S. Selick (Editor-in-Chief), Editors,** Stable and unstable homotopy, 1998

18 **Panos M. Pardalos and Henry Wolkowicz, Editors,** Topics in semidefinite and interior-point methods, 1998

17 **Joachim J. R. Cuntz and Masoud Khalkhali, Editors,** Cyclic cohomology and noncommutative geometry, 1997

16 **Victor P. Snaith, Editor,** Algebraic K-theory, 1997

15 **Stephen P. Braham, Jack D. Gegenberg, and Robert J. McKellar, Editors,** Sixth Canadian conference on general relativity and relativistic astrophysics, 1997

14 **Mourad E. H. Ismail, David R. Masson, and Mizan Rahman, Editors,** Special functions, q-series and related topics, 1997

13 **Peter A. Fillmore and James A. Mingo, Editors,** Operator algebras and their applications, 1997

12 **Dan-Virgil Voiculescu, Editor,** Free probability theory, 1997

11 **Colleen D. Cutler and Daniel T. Kaplan, Editors,** Nonlinear dynamics and time series: Building a bridge between the natural and statistical sciences, 1997

10 **Jerrold E. Marsden, George W. Patrick, and William F. Shadwick, Editors,** Integration algorithms and classical mechanics, 1996

9 **W. H. Kliemann, W. F. Langford, and N. S. Namachchivaya, Editors,** Nonlinear dynamics and stochastic mechanics, 1996

8 **Larry M. Bates and David L. Rod, Editors,** Conservative systems and quantum chaos, 1996

7 **William F. Shadwick, Perinkulam Sambamurthy Krishnaprasad, and Tudor Stefan Ratiu, Editors,** Mechanics day, 1996

6 **Anna T. Lawniczak and Raymond Kapral, Editors,** Pattern formation and lattice gas automata, 1996

5 **John Chadam, Martin Golubitsky, William Langford, and Brian Wetton, Editors,** Pattern formation: Symmetry methods and applications, 1996

4 **William F. Langford and Wayne Nagata, Editors,** Normal forms and homoclinic chaos, 1995

3 **Anthony Bloch, Editor,** Hamiltonian and gradient flows, algorithms and control, 1994

2 **K. A. Morris, Editor,** Control of flexible structures, 1993

1 **Michael J. Enos, Editor,** Dynamics and control of mechanical systems: The falling cat and related problems, 1993

Hydrodynamic Limits and Related Topics

FIELDS INSTITUTE COMMUNICATIONS

THE FIELDS INSTITUTE FOR RESEARCH IN MATHEMATICAL SCIENCES

Hydrodynamic Limits and Related Topics

Shui Feng
Anna T. Lawniczak
S. R. S. Varadhan
Editors

American Mathematical Society
Providence, Rhode Island

The Fields Institute for Research in Mathematical Sciences

The Fields Institute is named in honour of the Canadian mathematician John Charles Fields (1863–1932). Fields was a visionary who received many honours for his scientific work, including election to the Royal Society of Canada in 1909 and to the Royal Society of London in 1913. Among other accomplishments in the service of the international mathematics community, Fields was responsible for establishing the world's most prestigious prize for mathematics research—the Fields Medal.

The Fields Institute for Research in Mathematical Sciences is supported by grants from the Ontario Ministry of Education and Training and the Natural Sciences and Engineering Research Council of Canada. The Institute is sponsored by McMaster University, the University of Toronto, the University of Waterloo, and York University, and has affiliated universities in Ontario and across Canada.

2000 *Mathematics Subject Classification.* Primary 60K35, 82C22; Secondary 35B40, 35Q99, 37B15, 68Q80, 76X05, 82C05, 82C20, 82C21, 82C24.

Library of Congress Cataloging-in-Publication Data

Hydrodynamic limits and related topics / Shui Feng, Anna T. Lawniczak, S. R. S. Varadhan, editors.

p. cm. — (Fields Institute Communications, ISSN 1069-5265 ; v. 27)

Includes bibliographical references.

ISBN 0-8218-1993-3 (alk. paper)

1. Percolation (Statistical physics)—Congresses. I. Feng, Shui, 1962– II. Lawniczak, Anna T., 1953– III. Varadhan, S. R. S. IV. Series.

QC174.85.P45 H94 2000

530.13—dc21 00-063944

∞ The paper used in this book is acid-free and falls within the guidelines
established to ensure permanence and durability.
This publication was prepared by The Fields Institute.
Visit the AMS home page at URL: http://www.ams.org/

10 9 8 7 6 5 4 3 2 1 05 04 03 02 01 00

Contents

Preface

A workshop on "Hydrodynamic Limits" was held on October 7-10, 1998 at The Fields Institute for Research in Mathematical Sciences in Toronto, Canada. Preceding the workshop Prof. S.R.S. Varadhan delivered a mini-course consisting of six lectures on "Hydrodynamic Limits and Large Deviations". These were all part of the Institute's special year program on Probability and Its Applications.

The first part of this volume contains the lecture notes of the mini-course, and the second part includes refereed papers by the speakers and contributors to the workshop. The mini-course provides a comprehensive introduction to the theory of hydrodynamic limits including the entropy argument, relative entropy method, and the techniques for non-gradient systems. The articles appearing in part two reflect a number of research directions that are currently being pursued in hydrodynamic limits and related fields.

We would like to thank The Fields Institute for Research in Mathematical Sciences for its hospitality and financial support. We would also like to thank Jose A. Carrillo, Andreas Deutsch, Gregory Eyink, Henryk Fukś, Tadahisa Funaki, Jürgen Gärtner, Reinhard Illner, David Levermore, Jeremy Quastel, Krishnamur Ravishankar, Fraydon Rezakhanlou, Timo Seppäläinen, Herbert Spohn, and Horng-Tzer Yau for speaking at the workshop. We are grateful to all the contributors to this volume. Finally, we were fortunate to benefit from the help of Alesia Zuccala and the typing and editing job of Debbie Iscoe.

Shui Feng
Anna T. Lawniczak
Srinivasa R. S. Varadhan

Editors

July 2000

Part 1

Fields Institute Communications
Volume **27**, 2000

Lectures on Hydrodynamic Scaling

S.R.S. Varadhan
Courant Institute of Mathematical Sciences
New York University
251 Mercer Street
New York, NY 10012
varadhan@cims.nyu.edu

Abstract. Examples of hydrodynamic scaling, illustrating different methods including the relative entropy method are given. The use of the Dirichlet form as well as techniques from the theory of large deviations are demonstrated. Gradient and nongradient examples are considered.

1 From classical mechanics to Euler equations

The basic example of hydrodynamical scaling is naturally hydrodynamics itself. Let us start with a collection of $N \simeq \bar{\rho}\ell^3$ classical particles in a large periodic cube Λ_ℓ of side ℓ in R^3. The motion of the particles are governed by the equations of motion of a classical Hamiltonian dynamical system with energy given by

$$H(p\,,q) = \frac{1}{2}\sum_{i=1}^{N} \|p_i\|^2 + \frac{1}{2}\sum_{i\neq j} V(q_i - q_j)$$

Here, $q_i \in \Lambda_\ell$ is the position of the i-th particle and $p_i \in R^3$ is its velocity. The coordinates $k = 1, 2, 3$ refer to the three components of position or velocity. The repulsive potential $V \geq 0$ is an even function and has compact support in R^d. The interaction in particular is short range. The equations of motion are

$$\begin{aligned} \frac{dq_i^k}{dt} &= \frac{\partial H(p\,,q)}{\partial p_i^k} = p_i^k \\ \frac{dp_i^k}{dt} &= -\frac{\partial H(p\,,q)}{\partial q_i^k} = -\sum_{j=1}^{N} V_k(q_i - q_j) \end{aligned} \tag{1.1}$$

where

$$V_k(q) = \frac{\partial V(q)}{\partial q^k}\,; \quad k = 1, 2, 3$$

are the three components of the gradient of V. The dynamical system has five conserved quantities. The total number N of particles, the total momenta $\sum_{i=1}^{N} p_i^k$

2000 *Mathematics Subject Classification.* Primary 60K35; Secondary 82C22.

for $k = 1,2,3$ and the total energy $H(p,q)$. The hydrodynamic scaling in this context consists of rescaling space and time by a factor of ℓ. The rescaled space is the unit torus $\mathbf{T}^3$ in 3-dimensions. The macroscopic quantities to be studied correspond to the five conserved quantities. The first one of these is the density, and is measured by a function $\rho(t,x)$ of t and x. For each $\ell < \infty$ it is approximated by $\rho_\ell(t,x)$, defined by

$$\int_{\mathbf{T}^d} J(x)\rho_\ell(t,x)dx = \frac{1}{\ell^3}\sum_{i=1}^{N} J(\frac{q_i(\ell t)}{\ell}) \tag{1.2}$$

A straight forward differentiation with respect to t yields

$$\begin{aligned}\frac{d}{dt}\int_{\mathbf{T}^d} J(x)\rho_\ell(t,x)dx &= \frac{d}{dt}\frac{1}{\ell^3}\sum_{i=1}^{N} J(\frac{q_i(\ell t)}{\ell})\\ &= \frac{1}{\ell^3}\sum_{i=1}^{N}(\nabla J)(\frac{q_i(\ell t)}{\ell})\cdot p_i(\ell t)\\ &\simeq \int_{\mathbf{T}^d}(\nabla J)(x)\cdot \rho_\ell(t,x)u_\ell(t,x)dx\end{aligned} \tag{1.3}$$

where $u_\ell(t,x) = u^k_\ell(t,x)$, $k = 1,2,3$ are the components of the 'average' velocity of the fluid at the rescaled space time point x,t. This introduces three other macroscopic variables, which represent the three coordinates of the momenta that are conserved.

We can now write down the first of our five equations

$$\frac{\partial\rho}{\partial t} + \nabla\cdot(\rho u) = 0 \tag{1.4}$$

To derive the next three equations, using a test functions J, we differentiate for $k = 1,2,3$

$$\begin{aligned}\frac{d}{dt}\frac{1}{\ell^3}\sum_{i=1}^{N} J(\frac{q_i(\ell t)}{\ell})p^k_i(\ell t) &= \frac{1}{\ell^3}\sum_{i=1}^{N} p^k_i(\ell t)(\nabla J)(\frac{q_i(\ell t)}{\ell})\cdot p_i(\ell t)\\ &\quad - \frac{1}{\ell^2}\sum_{i=1}^{N}\sum_{j=1}^{N} J(\frac{q_i(\ell t)}{\ell})V_k(q_i(\ell t) - q_j(\ell t))\end{aligned} \tag{1.5}$$

If we now use the skew-symmetry of $V_k = \frac{\partial V}{\partial q_k}$, we can rewrite the second term of the right hand side of (1.5) as

$$\begin{aligned}&-\frac{1}{2\ell^2}\sum_{i=1}^{N}\sum_{j=1}^{N}(J(\frac{q_i(\ell t)}{\ell}) - J(\frac{q_j(\ell t)}{\ell})V_k(q_i(\ell t) - q_j(\ell t))\\ &\simeq -\frac{1}{2\ell^3}\sum_{i=1}^{N}\sum_{j=1}^{N} J_r(\frac{q_i(\ell t)}{\ell})(q^r_i(\ell t) - q^r_j(\ell t))V_k(q_i(\ell t) - q_j(\ell t))\\ &= \frac{1}{\ell^3}\sum_{i=1}^{N}\sum_{j=1}^{N} J_r(\frac{q_i(\ell t)}{\ell})\psi^r_k(q_i(\ell t) - q_j(\ell t))\end{aligned} \tag{1.6}$$

with

$$\psi_k^r(q) = -\frac{1}{2} q^r V_k(q)$$

The next step is rather mysterious and requires considerable explanation. The quantities

$$\sum_{i=1}^{N} p_i^k p_i^r, \quad \sum_{i,j=1}^{N} \psi_k^r(q_i(t) - q_j(t))$$

are not conserved. They depend on combinations of individual velocities that are not conserved and on spacings between particles both of which change in the microscopic time scale and therefore do so rapidly in the macroscopic time scale. They can therefore be replaced by their space-time averages. By appealing to an 'Ergodic Theorem' they can be replaced by their averages with repect to their equilibrium distributions. The equilibrium 'ensemble' consists of an infinite collection of points $\{p_\alpha, q_\alpha\}$, in the phase space $R^3 \times R^3$. There is a natural five parameter family of measures $\mu_{\rho,u,T}$ that are invariant under spatial translations as well as the Hamiltonian dynamics. The points $\{p_\alpha\}$ are distributed according to a Gibbs Distribution with density ρ and formal interaction energy

$$\frac{1}{2T} \sum_{\alpha,\beta} V(q_\alpha - q_\beta)$$

In other words $\{q_\alpha\}$ is a point process obtained by taking infinite volume limit of $N = \ell^3 \rho$ particles distributed in the cube of side ℓ in R^3 according to the joint density

$$\frac{1}{Z} \exp\left[-\frac{1}{2T} \sum_{1 \le i \ne j \le N} V(q_i - q_j)\right]$$

where Z is the normalization constant. The velocities $\{p_\alpha\}$ are distributed independently of each other as well as of $\{q_\alpha\}$, having a common 3-dimensional Gaussian distribution with mean u and covariance TI. Assuming that the infinite volume limit exists in a reasonable sense it will be a point process defined as an infinite volume Gibbs measure $\mu_{\rho,T}$. The velocities $\{p_\alpha\}$ will be an independent Gaussian ensemble $\nu_{u,T}$.

In the first term the quantities $p_i^k p_i^r$ are replaced by their expectations

$$u^k(t,x) u^r(t,x) + \delta_{k,r} T(t,x)$$

and in the second term $\psi_{k,r}$ are replaced by their expectations that involve the 'pressure' per unit volume in the Gibbs ensemble

$$\mathbf{P}_k^r(\rho, T) = \lim_{\ell \to \infty} E^{\mu_{\rho,T}} \left\{ \frac{1}{\ell^3} \sum_{\substack{|q_\alpha| \le \ell \\ |q_\beta| \le \ell}} \psi_k^r(q_\alpha - q_\beta) \right\}$$

This leads to the equation

$$\begin{aligned} \frac{d}{dt} \int_{\mathbf{T}^d} J(x) u^k(t,x) dx = \int_{\mathbf{T}^d} \sum_{r=1}^{3} \frac{\partial J}{\partial x_r}(x).(u^k(t,x) u^r(t,x) + \delta_{k,r} T(t,x)) dx \\ + \int_{\mathbf{T}^d} \sum_{r=1}^{3} \frac{\partial J}{\partial x_r}(x) \mathbf{P}_k^r(\rho(t,x), T(t,x)) dx \end{aligned} \tag{1.7}$$

We now integrate by parts, remove the test function J and obtain from (1.7)

$$\frac{d}{dt}(\rho u) + \nabla\cdot(\rho\, u\otimes u + \rho\, T I + \mathbf{P}(\rho\,,T)) = 0 \tag{1.8}$$

There is an equation of state that expresses the total energy per unit volume e as

$$e(\rho\,,u\,,T) = \frac{1}{2}\rho(|u|^2 + 3T) + f(\rho\,,T) \tag{1.9}$$

where $f(\rho\,,T)$, the potential energy per unit volume, is given by

$$f(\rho\,,T) = \lim_{\ell\to\infty} E^{\mu_{\rho,T}}\left\{\frac{1}{2\ell^3}\sum_{\substack{|q_\alpha|\le\ell\\ |q_\beta|\le\ell}} V(q_\alpha - q_\beta)\right\}$$

Although we will not derive it, there is a similar equation for $e(t,x)$ that is obtained by differentiating

$$\frac{d}{dt}\frac{1}{2\ell^3}\sum_{i=1}^{N} J(\frac{q_i(\ell t)}{\ell})\left[|p_i(\ell t)|^2 + \sum_{j=1}^{N} V(q_i(\ell t) - q_j(\ell t))\right]$$

and proceeding in a similar fashion. It looks like

$$\frac{de}{dt} + \nabla\cdot\left[(e+T)u + \mathbf{P}(\rho\,,T)u\right] = 0 \tag{1.10}$$

The five equations one for density given by (1.4), the three for velocities contained in (1.8) and finally the enrgy equation (1.10) constitute a first order system of non-linear hyperbolic conservation laws in the six variables $[\rho, u, T, e\,]$ with one relation between them given by (1.9). Given smooth initial data they have local solutions.

Rigorous derivation of these equations does not exist. The ergodic theory is definitely plausible, but hard to establish. The reason is that we have essentially an infinite system of ordinary differential equations representing a classical deterministic dynamical system and the ergodicity properties are nearly impossible to establish in any general context. However if we had some noise in the system, i.e. stochastic dynamics instead of deterministic dynamics, then we will be concerned with the ergodic theory of Markov Processes of some kind, which is far more accessible. This is essentially the focus of our lectures.

2 Some examples

The simplest example is a system of noninteracting particles undergoing independent motions. For example we could have on $\mathbf{T}^3$, $L \simeq \bar\rho N^3$ particles all behaving like independent Brownian Particles. If the initial configuration of the L particles is such that the empirical distribution

$$\nu_0(dx) = \frac{1}{N^3}\sum_i \delta_{x_i}$$

has a deterministic limit $\rho_0(x)dx$, then the empirical distribution

$$\nu_t(dx) = \frac{1}{N^3}\sum_i \delta_{x_i(t)}$$

of the configuration at time t, has a deterministic limit $\rho(t,x)dx$ as $N \to \infty$ and $\rho(t,x)$ can be obtained from $\rho_0(x)$ by solving the heat equation

$$\frac{\partial \rho}{\partial t} = \frac{1}{2}\Delta\rho$$

with the initial condition $\rho(0,x) = \rho_0(x)$. The proof is an elementary law of large numbers argument involving a calculation of two moments. Let $f(x)$ be a continuous function on $\mathbf{T}^3$ and let us calculate for

$$U = \frac{1}{N^3}\sum_i f(x_i(t)) \tag{2.1}$$

the first two moments given the initial configuration $(x_1, \cdots, x_L)$.

$$E(U) = \frac{1}{N^3}\sum \int_{\mathbf{T}^3} f(y)p(t,x_i,y)dy$$

and an elementary calculation reveals that the expectation converges to the constant:

$$\int_{\mathbf{T}^3}\int_{\mathbf{T}^3} f(y)p(t,x,y)\rho_0(x)dydx = \int_{\mathbf{T}^3} f(y)\rho(t,y)dy \tag{2.2}$$

Although the summamnds in (2.1) are not identically distributed, the independence clearly provides a uniform upper bound of the order CN^{-3} for the variance and goes to 0. Of course on $\mathbf{T}^3$ we could have had a process obtained by rescaling a random walk on a large torus of size N. Then the hydrodynamic scaling limit would be a consequence of the central limit theorem for a single particle under space time rescaling. Our result on the behavior of the empirical distribution of the particles will continue to be a consequence of the law of large numbers resulting from the averaging over L independently moving particles. The situation could be different if the particles interacted with each other.

The next class of examples are called simple exclusion processes. They make sense on any finite or countable set X and for us it will be either the integer lattice $\mathbf{Z}^d$ in d-dimensions or $\mathbf{Z}^d_N$ obtained from it as a quotient by considering each coordinate modulo N. At any given time a subset of these lattice sites will be occupied by particles, with atmost one particle at each site. In other words some sites are empty while others are occupied with one particle. The particles move randomly. Each particle waits for an exponential random time and then tries to jump from its current site x to a new site y. The new site y is picked randomly according to a probability distribution $\pi(x,y)$. In particular $\sum_y \pi(x,y) = 1$ for every x. Of course a jump to y is not always possible. If the site is empty the jump is possible and is carried out. If the site already has a particle, the jump cannot be carried out and the particle forgets about it and waits for another chance, i.e. waits for a new exponential waiting time.

If we normalize so that all waiting times have mean 1, the generator of the process can be written down as

$$(\mathcal{A}f)(\eta) = \sum_{x,y} \eta(x)(1-\eta(y))\pi(x,y)[f(\eta^{x,y}) - f(\eta)] \tag{2.3}$$

where η represents the configuration with $\eta(x) = 1$ if there is a particle at x and $\eta(x) = 0$ otherwise. For each configuration η and a pair of sites x, y the new configuration $\eta^{x,y}$ is defined by

$$\eta^{x,y}(z) = \begin{cases} \eta(y) & \text{if } z = x \\ \eta(x) & \text{if } z = y \\ \eta(z) & \text{if } z \neq x\,, y \end{cases} \tag{2.4}$$

We will be mainly concerned with the situation where the set X is $\mathbf{Z}^d$ or $\mathbf{Z}^d_N$, viewed naturally as an Abelian group with $\pi(x\,,y)$ being translation invariant and given by $\pi(x\,,y) = p(y-x)$ for some peobability distribution p. It is convenient to assume that p has finite support. There are various possibilities.

$$p \text{ is symmetric i.e. } p(x) = p(-x)$$

or more generally

$$p \text{ has mean zero i.e. } \sum_y yp(y) = 0$$

and finally

$$\sum_y y\,p(y) = m \neq 0$$

We shall first concentrate on the symmetric case. Let us look at the function

$$V_J(\eta) = \sum J(x)\eta(x)$$

and compute

$$\begin{aligned} (\mathcal{A}V_J)(\eta) &= \sum_{x,y} \eta(x)(1-\eta(y))p(y-x)(J(y)-J(x)) \\ &= \sum_{x,y} \eta(x)p(y-x)(J(y)-J(x)) \\ &= \sum_{x,y} \eta(x)[(\mathbf{P}-I)J](x) \\ &= V_{(\mathbf{P}-I)J}(\eta) \end{aligned} \tag{2.5}$$

The space of linear functionals is left invariant by the generator. It is not difficult to see that

$$E_\eta\big[V_J(\eta(t))\big] = V_{J(t)}(\eta)$$

where

$$J(t) = \exp[t(\mathbf{P}-I)]J$$

is the solution of

$$\frac{d}{dt}J(t\,,x) = (\mathbf{P}-I)J(t\,,x)$$

It is almost as if the interaction has no effect and in fact for the calculation of expectations of 'one particle' functions it clearly does not. Let us start with a configuration on $\mathbf{Z}^d_N$ and scale space by N and time by N^2. The generator becomes $N^2\mathcal{A}$ and the particles can be visualized as moving on a lattice imbedded in the unit torus $\mathbf{T}^d$, with a spacing of $\frac{1}{N}$, and becoming dense as $N \to \infty$.

Let us consider the functional

$$\xi(t) = \frac{1}{N^d}\sum_x J(\frac{x}{N})\eta_t(x)$$

and we can write

$$\xi(t) - \xi(0) = \int_0^t V_N(\eta(s))ds + M_N(t)$$

where

$$V_N(\eta) = (N^2 \mathcal{A} V_J)(\eta) = V_{J_N}(\eta)$$

with

$$(J_N)(x) = N^2 \sum \big[J(x + \frac{z}{N}) - J(x)\big]p(z)$$
$$\simeq \frac{1}{2}(\Delta_C J)(x)$$

Here Δ_C refers to the Laplacian

$$\sum_{i,j} C_{i,j} \frac{\partial^2}{\partial x_i \partial x_j}$$

with the covariance matrix C given by

$$C_{i,j} = \sum_x x_i x_j p(x)$$

$M_N(t)$ is a martingale and a very elementary calculation yields

$$E\Big\{[M_N(t)]^2\Big\} \leq C\, t\, N^{-d}$$

essentially completing the proof in this case. Technically the empirical distribution $\nu_N(t)$ is viewed as a measure on $\mathbf{T}^d$ and $\nu_N(\cdot)$ is viewed as a stochastic process with values in the space $\mathcal{M}(\mathbf{T}^d)$ of nonnegative measures on $\mathbf{T}^d$. In the limit it lives on the set of weak solutions of the heat equation

$$\frac{\partial \rho}{\partial t} = \frac{1}{2} \Delta_C\, \rho \tag{2.6}$$

and the uniqueness of such weak soultions for given initial density establishes the validity of the hydrodynamic limit. We could have have computed two moments as in the noninteracting case. The expectation would have been no diferent from the noninteracting case since it involves only one particle functions. The variance involves two particle functions and would have involved slighlty more work, because the independence is not there. The martingale argument however is more general.

Let us make the problem slightly more complicated by adding a small bias. Let $q(x)$ be an odd function with $q(-x) = -q(x)$ and we will modify the problem by making p depend on N in the form

$$p_N(x) = p(x) + \frac{1}{N} q(x) \tag{2.7}$$

Assuming that q is nonzero only when p is so, p_N will be an admissible transition probability for large enough N. A calculation yields that in the slightly modified model referred to as weakly asymmetric simple exclusion model V_N is given by

$$V_N(\eta) \simeq V_{J_N}(\eta) + \frac{1}{N^d} \sum \eta(x)(1 - \eta_x) < m\,, \nabla J(x) > \tag{2.8}$$

with

$$m = N \sum_x x\, p_N(x) = \sum_x x\, q(x)$$

If one thinks of $\rho(t,x)$ as the density of particles at the (macroscopic) time t and space x the first term on the right hand side of (2.8) clearly wants to have the limit

$$\int_{\mathbf{T}^d} \frac{1}{2}(\Delta_C J)(x)\rho(t,x)dx$$

It is not so clear what to do with the second term. The 'invariant' measures in this model are the Bernoulli measures with various densities ρ and upon 'averaging' the second term should converge to

$$\int_{\mathbf{T}^d} < m,(\nabla J)(x) > \rho(t,x)(1-\rho(t,x)dx$$

replacing the linear heat equation (2.6) by the nonlinear equation

$$\frac{\partial \rho}{\partial t} = \frac{1}{2}\Delta_C \rho - \nabla \cdot m\rho(1-\rho) \tag{2.9}$$

This requires justification and that will be the content of our next lecture.

Let us now turn to the case where p has mean zero but is not symmetric. In this case

$$V_N(\eta) = N^{2-d}\sum_{x,y}\eta(x)(1-\eta(y))p(y-x)[J(\frac{y}{N}) - J(\frac{x}{N})] \tag{2.10}$$

and we get stuck at this point. If p is symmetric, as we saw, we gain a factor of N^{-2}. Otherwise the gain is only a factor of N^{-1} which is not enough. We seem to end up with

$$\begin{aligned} N^{-d}&\sum_x\sum_y \eta(x) < \frac{1}{2}[(\nabla J)(\frac{x}{N}) + (\nabla J)(\frac{y}{N})], N(1-\eta(y))(y-x)p(y-x) > \\ &= \frac{1}{2N^d}\sum_x (\nabla J)(\frac{x}{N})N\Psi_x \end{aligned} \tag{2.11}$$

where

$$\begin{aligned} \Psi_0 &= \frac{1}{2}[\eta(0)\sum_z(1-\eta(z))z\,p(z) + (1-\eta(0))\sum_z \eta(-z)z\,p(z)] \\ &= \frac{1}{2}[-\eta(0)\sum_z \eta(z)z\,p(z) + (1-\eta(0))\sum_z \eta(-z)z\,p(z)] \\ &= \frac{1}{2}[\sum_z \eta(-z)z\,p(z) - \eta(0)\sum_z(\eta(z)+\eta(-z))z\,p(z)] \end{aligned} \tag{2.12}$$

The second sum is zero in the symmetric case and Ψ_0 can then be written as a 'gradient'

$$\Psi_0 = \sum_j \tau_{e_j}\xi_j - \xi_j$$

where τ_{e_j} are shifts in the coordinate directions. This allows us to do summation by parts and gain a factor of N^{-1}. When this is not the case, we have a 'nongradient' model and the hydrodynamic limit can no longer be established by simple averaging.

We will leave this somewhat difficult case till the end for lectures 5 and 6 but take up the third case

$$\sum_z zp(z) = m \neq 0$$

that needs hyperbolic scaling and leads to Burgers equation with zero viscosity in lecture 4.

3 Gradient models, Dirichlet forms and large deviations

The important ingredient in the analysis of gradient models is the ability to do averaging and replace quantities by their expected values calculated under various equilibrium distributions. Suppose $\mu_N$is a probability measure on the space Ω_N of configurations η on the periodic lattice $\mathbf{Z}_N^d$. We wish to think of μ_N as being a Bernoulli measure with some density ρ. The density ρ is not quite a constant, but a slowly varying function on $\mathbf{Z}_N^d$, and in fact a function of the macroscopic variable $\frac{x}{N}$. Let $g = g(\eta)$ be a local function depending on the configuration in some finite box around 0. By Av_ℓ we will denote the averaging process over the block(cube) $B_\ell = \{z : |z_j| \leq \ell; 1 \leq j \leq d\}$ of side $2\ell + 1$

$$(Av_\ell g)(\eta) - \frac{1}{(2\ell+1)^d} \sum_{z \in B_\ell} g(\tau_z \eta) \tag{3.1}$$

If $g(\eta) = \eta(0)$ this produces the empirical density $\bar{\eta}_{\ell,0}$. For any local g we can calculate its expected value with repect to the Bernoulli measure with density ρ and get

$$\hat{g}(\rho) = E^{P_\rho}[g(\eta)] \tag{3.2}$$

The validity of the averaging principle means that we can replace $(Av_\ell g)(\eta)$ by $\hat{g}(\bar{\eta}_{\ell,0})$. More generally if we denote by τ_x the translation by x on the lattice then we can replace $(Av_\ell g)(\tau_y \eta)$ by $\hat{g}(\bar{\eta}_{\ell,y})$ or more precisely showing that the error

$$\delta_{N,\ell} = E^{\mu_N}\left[\frac{1}{N^d}\sum_x |(Av_\ell g)(\tau_x \eta) - \hat{g}(\bar{\eta}_{\ell,x})|\right] \tag{3.3}$$

goes to zero as $N \to \infty$ so long as $\ell \to \infty$ and $\frac{\ell}{N} \to 0$. What property of the sequence μ_N will allow us to make such a conclusion?

Dirichlet Form. Given the density $\{f(\eta)\}$ with respect to the uniform distribution of a probability distribution μ on $\mathbf{Z}_N^d$, the Dirichlet form $\mathcal{D}_N(\mu)$ is the quantity

$$\mathcal{D}_N(\mu) = \frac{1}{2^{N^d}} \sum_\eta \sum_{\substack{x,y \in \mathbf{Z}_N^d \\ |x-y|=1}} [\sqrt{f(\eta^{x,y})} - \sqrt{f(\eta)}|^2 \tag{3.4}$$

Remark: One should think of the inner summation as being over all the nearest neighbor bonds. The outer summation with the factor $\frac{1}{2^{N^d}}$ is integration with respect to the uniform distribution on Ω_N. Since f is a nonnegative L_1 function, its square root will be naturally an L_2 function and we are really dealing with the Dirichlet form of $\sqrt{f}$.

Theorem 3.1 (Averaging principle) *If μ_N on Ω_N is a sequence of probability distributions such that*

$$\mathcal{D}_N(\mu_N) \leq CN^{d-2} \tag{3.5}$$

for some constant C independent of N, then for every local g

$$\lim_{\epsilon\to 0}\limsup_{N\to\infty} \delta_{N,\epsilon N} = 0 \tag{3.6}$$

Proof The proof is carried out in two steps. First we show that

$$\lim_{\ell\to\infty}\limsup_{N\to\infty} \delta_{N,\ell} = 0 \tag{3.7}$$

and then

$$\lim_{\substack{\epsilon\to 0\\ \ell\to\infty}} E^{\mu_N}\left[\frac{1}{N^d}\sum_{x\in \mathbf{Z}_N^d} |\bar\eta_{\ell,x} - \bar\eta_{N\epsilon,x}|\right] = 0 \tag{3.8}$$

Being a polynomial in ρ, $\hat g(\rho)$ is a continuos function of ρ and (3.6) will follow from (3.7) and (3.8). □

Step 1: Since the details of the proof can be found in [KOV] we will only sketch it here. Suppose B_ℓ is a block of size $2\ell+1$ and $q_\ell(\eta)$ is an assignment of probabilities for the $2^{(2\ell+1)^d}$ possible configurations. We view the block non-periodically and we have $(2\ell)^d$ interior nearest neighbor bonds. The Dirichlet form is given by

$$\mathcal{D}_\ell(q_\ell) = \sum_{\substack{x,y\in B_\ell\\ |x-y|=1}} |\sqrt{q_\ell(\eta^{x,y})} - \sqrt{q_\ell(\eta)}|^2$$

Given a probability distribution μ_N on Ω_N, we denote its marginal on a block of size $2\ell+1$ centered at x by $\nu_{\ell,N,x}$ and by $\bar\nu_{\ell,N}$ we denote the average

$$\bar\nu_{\ell,N} = \frac{1}{N^d}\sum_{x\in\mathbf{Z}_N^d} \nu_{\ell,N,x}$$

over all the sites x. If we denote by $q_{\ell,N}(\eta)$ the individual probabilities the following inequality is a consequence of the convexity of $\mathcal{D}(\nu)$ in ν

$$\mathcal{D}_\ell(q_{\ell,N}) \leq (\frac{2\ell}{N})^d \mathcal{D}_N(\mu_N) \leq (\frac{2\ell}{N})^d CN^{d-2} = \frac{C(2\ell)^d}{N^2} \tag{3.9}$$

Let us denote the set of limit points of $q_{\ell,N}$ as $N\to\infty$ by $\mathcal{M}_\ell$. For any $\nu\in\mathcal{M}_\ell$ the associated probability function q satisfies $\mathcal{D}_\ell(q)=0$. Therefore it is invariant under exchanges between neighboring sites, and consequently under all permutations. The conditional distribution under ν of occupied sites, given the total number of particles k, is therefore the uniform distribution over all subsets of cardinality k. If we denote this uniform distribution by $\lambda_{\ell,k}$, then every $\nu\in\mathcal{M}_\ell$ is of the form

$$\nu = \sum_k \lambda_{\ell,k}\pi_\nu(k) \tag{3.10}$$

where $\pi_\nu(k)$ is the probability that under ν the total number of particles in the block of size $(2\ell+1)^d$ is equal to k. Since $\lambda_{\ell,k}$ converges to the Bernoulli with density ρ, if $\frac{k}{(2\ell+1)^d} \to \rho$ it follows that

$$\begin{aligned}
&\lim_{\ell\to\infty} \sup_{\nu\in\mathcal{M}_\ell} E^{\nu}[|Av_\ell\, g(\eta) - \hat{g}(Av_\ell\, \eta)|] \\
&\qquad \leq \lim_{\ell\to\infty} \sup_{0\leq k\leq(2\ell+1)^d} E^{\lambda_{k,\ell}}[|Av_\ell\, g(\eta) - \hat{g}(\frac{k}{(2\ell+1)^d})|] \qquad (3.11)\\
&\qquad = 0
\end{aligned}$$

establishing (3.7).

Step 2: For step 2, we need only to establish that the local density $\bar{\eta}_{\ell,x}$ does not fluctuate over small macroscopic length scales. This will follow from the estimate

$$\limsup_{\substack{\ell\to\infty\\ \epsilon\to 0}} \limsup_{N\to\infty} \sup_{|y|\leq N\epsilon} E^{\mu_N}\left[\frac{1}{N^d}\sum_{x\in\mathbf{Z}_N^d} |\bar{\eta}_{\ell,x} - \bar{\eta}_{\ell,x+y}|\right] = 0 \qquad (3.12)$$

If we take two blocks of size $2\ell+1$ centered at x and $x+y$ there is no bond connecting x and $x+y$ which are far from each other. We can introduce a Dirichlet form for this bond.

$$\mathcal{D}_{x,x+y}(q) = \sum_{\eta} |\sqrt{q(\eta^{x,x+y})} - \sqrt{q(\eta)}|^2$$

and this can be estimated, by writing the difference as a telescopic sum over exchanges involving nearest neighbors and applying Schwartz's inequlaity, interms of Dirichlet forms for nearest neighbor bonds. If we denote by $\nu_{\ell,N,x,x+y}$ the joint marginal over two widely seperated blocks of size $2\ell+1$ centered at x and $x+y$, and by

$$\bar{\nu}_{\ell,N,y} = \frac{1}{N^d}\sum_{x\in\mathbf{Z}_N^d} \nu_{\ell,N,x,x+y}$$

the average over all rigid spatial translations of the pair of blocks

$$\mathcal{D}_{\ell,0,y}(\bar{\nu}_{\ell,N,y}) \leq C|y|^2 N^{-2} \leq C\epsilon^2$$

It follows, by pure thought, (once $\epsilon\to 0$, the two averages are essentially averages over two halves of a combined system of $2(2\ell+1)^d$ sites which is in equilibrium) that

$$\limsup_{\ell\to\infty} \limsup_{\epsilon\to 0} \limsup_{N\to\infty} \sup_{|y|\leq N\epsilon} E^{\mu_N}\left[\frac{1}{N^d}\sum_{x\in\mathbf{Z}_N^d} |\bar{\eta}_{\ell,x} - \bar{\eta}_{\ell,x+y}|\right] = 0$$

Because $\bar{\eta}_{N\epsilon,x}$ is not very different from the average $Av_{|y|\leq N\epsilon}\, \bar{\eta}_{\ell,x+y}$ (3.8) follows.

Superexponential Estimates: Let us denote by P_N the measure corresponding to the process on $D[[0,T];\Omega_N]$ starting from some initial distribution. We want to replace additive functionals of the form

$$\int_0^T \frac{1}{N^d}\sum_x J(\frac{x}{N}) g((\tau_x\eta)(s))ds$$

by

$$\int_0^T \frac{1}{N^d}\sum_x J(\frac{x}{N})\hat{g}(\bar{\eta}_{N\epsilon,x})(s))ds$$

and for the difference

$$F_N = \int_0^T \frac{1}{N^d}\sum_x J(\frac{x}{N})g((\tau_x\eta)(s))ds - \int_0^T \frac{1}{N^d}\sum_x J(\frac{x}{N})\hat{g}(\bar{\eta}_{N\epsilon,x})(s))ds$$

we obtain the following superexponential bound.

Theorem 3.2 *For any* $\delta > 0$

$$\limsup_{\epsilon\to 0}\limsup_{N\to\infty}\frac{1}{N^d}\log P_N[|F_N| \geq \delta] = -\infty$$

Proof It suffices to prove that for any $\lambda > 0$

$$\limsup_{\epsilon\to 0}\limsup_{N\to\infty}\frac{1}{N^d}\log E^{P_N}\Big[\exp\big[\lambda N^d F_N\big]\Big] = 0 \tag{3.13}$$

If we denote by $\bar{P}_N$ the process that is initialized to start from Bernoulli measure with density $\frac{1}{2}$ which is an invariant measure then the Radon-Nikodym derivative $\frac{dP_N}{d\bar{P}_N}$ is bounded by 2^{N^d}. Therefore

$$E^{P_N}\Big[\exp\big[\lambda N^d F_N\big]\Big] \leq 2^{N^d} E^{\bar{P}_N}\Big[\exp\big[\lambda N^d F_N\big]\Big] \tag{3.14}$$

If we can show that under equilibrium we have

$$\limsup_{\epsilon\to 0}\limsup_{N\to\infty}\frac{1}{N^d}\log E^{\bar{P}_N}\Big[\exp\big[\lambda N^d F_N\big]\Big] = 0 \tag{3.15}$$

then from (3.14)

$$\limsup_{\epsilon\to 0}\limsup_{N\to\infty}\frac{1}{N^d}\log E^{P_N}\Big[\exp\big[\lambda N^d F_N\big]\Big] \leq \log 2$$

By Hölder's inequality

$$\log E^{P_N}\Big[\exp\big[\lambda N^d F_N\big]\Big] \leq \frac{1}{a}\log E^{P_N}\Big[\exp\big[a\lambda N^d F_N\big]\Big] \leq \frac{\log 2}{a}$$

and since $a > 0$ can be made arbitrarily large we obtain (3.13)

By Feynamn-Kac (3.15) can be reduced to an estimation of an eigen value. More precisely we only need to show that for every $\lambda > 0$

$$\limsup_{\epsilon\to 0}\limsup_{N\to\infty}\frac{1}{N^d}\Big[\sup_\nu E^\nu\big[N^d\lambda F_N\big] - cN^2\mathcal{D}_N(\nu)\Big] = 0$$

or

$$\limsup_{\epsilon\to 0}\limsup_{N\to\infty}\Big[\sup_\nu E^\nu\big[\lambda F_N\big] - \frac{c}{N^{d-2}}\mathcal{D}_N(\nu)\Big] = 0 \tag{3.16}$$

The apperance of the constant c is because the actual Dirichlet form involves $p(\cdot)$ and if we assume irreducibility we can get a bound. The factor N^2 is due to the speed up of time. If F_N is bounded by A, we will only see ν with

$$\mathcal{D}_N(\nu) \leq \frac{\lambda A}{c} N^{d-2} = CN^{d-2}$$

(3.16) now follows from Theorem 3.1. This concludes the proof. □

Application 1: Weak asymmetry. If Q_N is the process with weak asymmetry then

$$R_N = \frac{dQ_N}{dP_N}$$

can be explicitly calculated by Girsanov type formula and

$$E^{P_N}\left[R_N^p\right] \leq \exp\left[CN^d\right]$$

Therefore Q_N inherits the super-exponential bounds. This establishes the hydrodynamic limit for weakly asymmetric perturbations.

In fact if we perturb $p(\cdot)$ by a skew symmetric $\frac{q}{N}$ with q of the form $q(t, \frac{x}{N}, \cdot)$ so that the generator looks like

$$\begin{aligned} N^2 \sum_{x,y} &\eta(x)(1-\eta(y))p(y-x)[f(\eta^{x,y}) - f(\eta)] \\ &+ N \sum_{x,y} \eta(x)(1-\eta(y))q(t, \frac{x}{N}, y-x)[f(\eta^{x,y}) - f(\eta)] \end{aligned}$$

and the means are calculated as

$$b(t,\theta) = \sum_z z\, q(t,\theta,z),$$

then the hydrodynamic limit is

$$\frac{\partial \rho}{\partial t} = \frac{1}{2}\nabla \cdot C\nabla\rho - \nabla \cdot b\rho(1-\rho) \tag{3.17}$$

Application 2: Large deviations. The Entropy cost of the perturbation is calculated easily by the use of Girsanov's formula

$$H(Q_N, P_N) = E^{Q_N}\left[\int_0^T \sum_{x,y} \left[\eta_t(x)(1-\eta_t(y))c_N(t, \frac{x}{N}, y-x)\right] dt\right]$$

where c is the relative entropy of one Poisson distribution with respect to another

$$\begin{aligned} C_N(t,\theta,z) &= (N^2 p(z) + Nq(t,\theta,z)) \log \frac{N^2 p(z) + Nq(t,\theta,z)}{N^2 p(z)} \\ &\quad - (N^2 p(z) + Nq(t,\theta,z)) + N^2 p(z) \\ &\simeq \frac{q^2(t,\theta,z)}{2p(z)} \end{aligned}$$

It is clearly in our interest to minimize

$$\frac{1}{2}\sum_z \frac{q^2(t,\theta,z)}{p(z)}$$

subject to

$$\sum_z z\, q(t,\theta,z) = b$$

and the minimum is seen to be

$$\frac{1}{2} < b\,, C^{-1}\, b >$$

so that the minimal entropy cost for the large deviation of the empirical density from the solution of the normal hydrodynamic limiting equation

$$\frac{\partial \rho}{\partial t} = \frac{1}{2} \nabla \cdot C \nabla \rho \tag{3.18}$$

to a ρ that is the solution of (3.17) is asymptotically

$$N^d \mathcal{E}(b) = \frac{1}{2} N^d \int_0^T \int_{\mathbf{T}^d} \rho(t\,,\theta)(1-\rho(t\,,\theta)) < b(t,\theta)\,, C^{-1} b(t,\,\theta) > dt d\theta$$

We have now a large deviation lower bound for a ρ not satisfying (3.18). Find the class $\mathcal{B}(\rho)$ of b's that satisfy (3.17) and optimize $\mathcal{E}(b)$ over $\mathcal{B}(\rho)$. In other words we have established the large deviation lower bound with the rate function

$$I(\rho) = \inf_{b \in \mathcal{B}(\rho)} \mathcal{E}(b) \tag{3.19}$$

We now work on the upper bound. We start with a class of exponential martingales. For any smooth test function $J(t,\theta)$

$$\begin{aligned} E^{P_N}\Bigg[\exp\Bigg[\sum_x J(T, \frac{x}{N})\, \eta_T(x) - \sum_x J(0, \frac{x}{N})\, \eta_0(x) - \int_0^T \sum_x J_t(t, \frac{x}{N}) \eta_t(x) dt \\ - N^2 \int_0^T \sum_{x,y} p(y-x) \eta_t(x)(1-\eta_t(y)) \big[\exp[J(t, \frac{y}{N}) - J(t, \frac{x}{N})] - 1 \big] \Bigg] dt \Bigg] = 1 \end{aligned}$$

The quantity

$$N^2 \sum_{x,y} p(y-x) \eta_t(x)(1-\eta_t(y)) \big[\exp[J(t, \frac{y}{N}) - J(t, \frac{x}{N})] - 1 \big]$$

simplifies to

$$\frac{1}{2} \sum_x (\nabla \cdot C \nabla J)(t, \frac{x}{N}) \eta_t(x) + \frac{1}{2} \sum_{x,y} p(y-x) \eta_t(x)(1-\eta_t(y)) \, | < y-x\,, (\nabla J)(t, \frac{x}{N}) > |^2$$

Because of super exponential bounds we can effectively average the second term, even for large deviation purposes and interms of the density ρ, the expression looks like

$$\begin{aligned} N^d \Bigg[\frac{1}{2} \int_0^T \int_{\mathbf{T}^d} \nabla \cdot C \nabla J(t,\theta) \rho(t,\theta) dt d\theta \\ + \frac{1}{2} \int_0^T \int_{\mathbf{T}^d} < \nabla J(t,\theta), C \nabla J(t,\theta) > \rho(t,\theta)(1-\rho(t,\theta)) dt d\theta \Bigg] \end{aligned}$$

Formally we are looking at

$$E^{P_N} \Big[\exp \big[N^d F_N \big] \Big] = 1$$

where

$$
\begin{aligned}
F_N &= F_{N,J}(\rho(\cdot,\cdot)) \\
&\simeq \int_{\mathbf{Z}^d} J(T,\theta)\rho(T,\theta)d\theta - \int_{\mathbf{Z}^d} J(0,\theta)\rho(0,\theta)d\theta - \int_0^T \int_{\mathbf{Z}^d} J_t(t,\theta)\rho(t,\theta)dtd\theta \\
&- \frac{1}{2}\int_0^T \int_{\mathbf{T}^d} \nabla\cdot C\nabla J(t,\theta)\rho(t,\theta)dtd\theta \\
&- \frac{1}{2}\int_0^T \int_{\mathbf{T}^d} < \nabla J(t,\theta), C\nabla J(t,\theta) > \rho(t,\theta)(1-\rho(t,\theta))dtd\theta
\end{aligned}
$$

By standard large deviation theory one gets an upper bound with a rate function

$$\sup_J F_{N,J}(\rho(\cdot,\cdot))$$

that is easily seen to equal (3.19).

4 Relative entropy method

We will look at the situation where $m = \sum_z z\,p(z) \neq 0$. In order to convey the idea it is sufficient to restrict ourselves to the case wher $d = 1$, and $p(1) = 1$ and $p(z) = 0$ for all $z \neq 0$. This is the totally asymmetric nearest neighbor simple exclusion model and of course $m = 1$. The method that we describe here can be found in [Y] where it is introduced in connection with a different model.

In this case the rescaling in time is done by a factor of N and the generator is

$$(\mathcal{A}_N f)(\eta) = N \sum_{x\in\mathbf{Z}_N} \eta(x)(1-\eta(x+1))[f(\eta^{x,x+1}) - f(\eta)] \tag{4.1}$$

We can easily calculate

$$
\begin{aligned}
&d[\frac{1}{N}\sum_{x\in\mathbf{Z}_N} J(\frac{x}{N})\eta_t(x)] \\
&\quad= [\sum_{x\in\mathbf{Z}_N} [J(\frac{x+1}{N}) - J(\frac{x}{N})]\eta_t(x)(1-\eta_t(x+1))]dt + M_N(t) \\
&\quad\simeq \frac{1}{N}[\sum_{x\in\mathbf{Z}_N} J'(\frac{x}{N})\eta_t(x)(1-\eta_t(x+1))]dt + o(1)
\end{aligned} \tag{4.2}
$$

The martingale term is negligible and if do simple 'averaging' we get the equation

$$\frac{d}{dt}\int_{\mathbf{T}} J(\theta)\rho(t,\theta)d\theta = \int_{\mathbf{T}} J'(\theta)\rho(t,\theta)(1-\rho(t,\theta))d\theta$$

or equivalently

$$\frac{\partial\rho}{\partial t} + \frac{\partial[\rho(1-\rho)]}{\partial\theta} = 0 \tag{4.3}$$

Let us consider a smooth solution of the above equation in $[0,T]\times\mathbf{T}$. Let us assume $c \le \rho(t,\theta) \le 1-c$ for some $c > 0$. On $\mathbf{Z}_N$ we start with an initial distribution α_N where $\eta(x)$ are independent under α_N with $\alpha_N[\eta(x) = 1] = \rho(0,\frac{x}{N})$. At any time $t \in [0,T]$ we have the actual distribution of the particle system obtained by solving the Kolmogorov forward equation

$$\frac{d\mu_N(t)}{dt} = \mathcal{A}_N^*\mu_N(t)$$

with $\mu_N(0) = \alpha_N$. We can define $\alpha_N(t)$ from $\rho(t\,,\cdot)$ the same way we defined α from $\rho(0\,,\cdot)$ at time 0. Although $\alpha_N(t)$ and $\mu_N(t)$ match by our choice at $t = 0$, in general they do not do so for $t > 0$. However we do have

Theorem 4.1 *The relative entropy $H(\mu_N(t)\,;\alpha_N(t))$ of $\mu_N(t)$ with respect to $\alpha_N(t)$ satisfies*

$$\lim_{N\to\infty}\sup_{0\le t\le T}\frac{H(\mu_N(t)\,;\alpha_N(t))}{N} = 0$$

By simple large deviation estimates this does imply the validity of the hydrodynamic scaling limit. In particular

$$\limsup_{N\to\infty}\sup_{o\le t\le T} E^{\mu_N(t)}\left|\frac{1}{N}\sum_x J(\frac{x}{N})\eta(x) - \int_{\mathbf{T}} J(\theta)\rho(t\,,\theta)d\theta\right| = 0$$

The proof involves careful reduction to Gronwall's lemma. The first step is to compute

$$\begin{aligned}
\frac{d}{dt}H(\mu_N(t)\,;\alpha_N(t)) &= \frac{d}{dt}\sum_\eta \mu_N(t\,,\eta)\log\frac{\mu_N(t\,,\eta)}{\alpha_N(t\,,\eta)}\\
&= \frac{d}{dt}\sum_\eta \mu_N(t\,,\eta)[\log\mu_N(t\,,\eta) - \log\alpha_N(t\,,\eta)]\\
&= \sum_\eta(\mathcal{A}_N^*\mu_N(t\,,\cdot))(\eta)[\log\mu_N(t\,,\eta) - \log\alpha_N(t\,,\eta)]\\
&\quad + \sum_\eta \mu_N(t\,,\eta)[\frac{\frac{d}{dt}\mu_N(t\,,\eta)}{\mu_N(t\,,\eta)} - \frac{d}{dt}\log\alpha_N(t\,,\eta)]\\
&= \sum_\eta \mu_N(t\,,\eta)[\mathcal{A}_N(\log\frac{\mu_N(t\,,\cdot)}{\alpha_N(t\,,\cdot)})](\eta)\\
&\quad - \sum_\eta \mu_N(t\,,\eta)(\frac{d}{dt}\log\alpha_N(t\,,\eta))\\
&\le \sum_\eta \mu_N(t\,,\eta)\left[\frac{[\mathcal{A}_N\frac{\mu_N(t\,,\cdot)}{\alpha_N(t\,,\cdot)}](\eta)}{\frac{\mu_N(t\,,\eta)}{\alpha_N(t\,,\eta)}}\right]\\
&\quad - \sum_\eta \mu_N(t\,,\eta)(\frac{d}{dt}\log\alpha_N(t\,,\eta))\\
&= \sum_\eta \alpha_N(t\,,\eta)[\mathcal{A}_N\frac{\mu_N(t\,,\cdot)}{\alpha_N(t\,,\cdot)}](\eta)\\
&\quad - \sum_\eta \mu_N(t\,,\eta)(\frac{d}{dt}\log\alpha_N(t\,,\eta))\\
&= \sum_\eta \mu_N(t\,,\eta)\frac{[(\mathcal{A}_N^* - \frac{d}{dt})\alpha_N(t\,,\cdot)](\eta)}{\alpha_N(t\,,\eta)}\\
&= E^{\mu_N(t)}\left[\frac{[(\mathcal{A}_N^* - \frac{d}{dt})\alpha_N(t\,,\cdot)](\eta)}{\alpha_N(t\,,\eta)}\right]
\end{aligned}$$

Explicitly

$$\begin{aligned}\alpha_N(t\,,\eta) &= \prod_x \rho(t\,,\frac{x}{N})^{\eta(x)}[1-\rho(t\,,\frac{x}{N})]^{1-\eta(x)}\\ &= \exp\sum_x \eta(x)\log\rho(t\,,\frac{x}{N}) + (1-\eta(x))\log(1-\rho(t\,,\frac{x}{N}))\\ &= \exp\psi_N(t\,,\eta)\end{aligned}$$

We compute

$$\frac{[\mathcal{A}_N^*\alpha_N(t\,,\cdot)](\eta)}{\alpha_N(t\,,\eta)} = N\sum_x(1-\eta(x))\eta(x+1)\big[\exp[\psi(t\,,\eta^{x,x+1}) - \psi(t\,,\eta)] - 1\big]$$

and

$$\begin{aligned}&\exp[\log\rho(t\,,\frac{x}{N}) + \log(1-\rho(t\,,\frac{x+1}{N})) - \log(1-\rho(t\,,\frac{x}{N})) - \log\rho(t\,,\frac{x+1}{N})] - 1\\ &= -\frac{1}{N}\frac{\rho'}{\rho}(t\,,\frac{x}{N}) - \frac{1}{N}\frac{\rho'}{1-\rho}(t\,,\frac{x}{N}) + o(\frac{1}{N})\end{aligned}$$

Therefore

$$\begin{aligned}\frac{[\mathcal{A}_N^*\alpha_N(t\,,\cdot)](\eta)}{\alpha_N(t\,,\eta)} &= -\sum_x(1-\eta(x))\eta(x+1)[\frac{\rho'}{\rho}(t\,,\frac{x}{N}) + \frac{\rho'}{1-\rho}(t\,,\frac{x}{N})] + o(N)\\ &= -\sum_x(1-\eta(x))\eta(x+1)\frac{\rho'}{\rho(1-\rho)}(t\,,\frac{x}{N})\end{aligned}$$

and

$$-\frac{d}{dt}\psi_n(t\,,x) = -\sum_x\eta(x)\frac{\rho_t}{\rho}(t\,,\frac{x}{N}) + \sum_x(1-\eta(x))\frac{\rho_t}{1-\rho}(t\,,\frac{x}{N})$$

There is local equilibrium over large microscopic blocks under $\mu_N(t)$. We can therefore replace $(1-\eta(x))\eta(x+1)$ by $\bar\eta_{k,x}(1-\bar\eta_{k,x})$. We get using $\rho_t = -[\rho(1-\rho)]' =$

$(2\rho-1)\rho'$

$$
\begin{aligned}
&\frac{d}{dt}\frac{H(\mu_N(t)\,;\alpha_N(t))}{N}\\
&\le E^{\mu_N(t)}\Big[\frac{1}{N}\sum_x \bar\eta_{k,x}(1-\bar\eta_{k,x})\frac{-\rho'}{\rho(1-\rho)}(t\,,\frac{x}{N})\\
&\qquad -\bar\eta_{k,x}\frac{(2\rho-1)\rho'}{\rho}(t\,,\frac{x}{N})+(1-\bar\eta_{k,x})\frac{(2\rho-1)\rho'}{1-\rho}(t\,,\frac{x}{N})\Big]+\epsilon_{k,N}\\
&= E^{\mu_N(t)}\Big[\frac{1}{N}\sum_x \bar\eta_{k,x}(1-\bar\eta_{k,x})\frac{-\rho'}{\rho(1-\rho)}(t\,,\frac{x}{N})\\
&\qquad -\bar\eta_{k,x}\frac{(2\rho-1)\rho'}{\rho}(t\,,\frac{x}{N})-\bar\eta_{k,x}\frac{(2\rho-1)\rho'}{1-\rho}(t\,,\frac{x}{N})\Big]+\epsilon_{k,N}\\
&= E^{\mu_N(t)}\Big[\frac{1}{N}\sum_x \bar\eta_{k,x}(1-\bar\eta_{k,x})\frac{-\rho'}{\rho(1-\rho)}(t\,,\frac{x}{N})\\
&\qquad -\bar\eta_{k,x}\frac{(2\rho-1)\rho'}{\rho(1-\rho)}(t\,,\frac{x}{N})\Big]+\epsilon_{k,N}\\
&\le \frac{1}{N}\log E^{\alpha_N(t)}\Big[\exp\big[\delta V_{N,k}(\eta)\big]\Big]+\frac{1}{\delta}H(\mu_N(t)\,;\alpha_N(t))+\epsilon_{k,N}
\end{aligned}
$$

where

$$
\begin{aligned}
V_{N,k}(\eta)&=\sum_x \bar\eta_{k,x}(1-\bar\eta_{k,x})\frac{-\rho'}{\rho(1-\rho)}(t\,,\frac{x}{N})-\bar\eta_{k,x}\frac{(2\rho-1)\rho'}{\rho(1-\rho)}(t\,,\frac{x}{N})\\
&=\sum_x \rho'(t\,,\frac{x}{N})F(\bar\eta_{k,x}\,,\rho(t\,,\frac{x}{N}))
\end{aligned}
$$

and

$$
F(m\,,\rho)=-m\frac{(2\rho-1)}{\rho(1-\rho)}-\frac{m(1-m)}{\rho(1-\rho)}
$$

The important fact is

$$
F_m(m\,,\rho)=-\frac{(2\rho-1)}{\rho(1-\rho)}-\frac{1-2m}{\rho(1-\rho)}=0
$$

if $\rho=m$. Therefore

$$
|F(m\,,\rho)-F(\rho\,,\rho)|\le C|m-\rho|^2
$$

Since we could easily subtract from $V_{N,k}(\eta)$ a sum of the type

$$
\sum_x \rho'(t\,,\frac{x}{N})F(\rho(t\,,\frac{x}{N})\,,\rho(t\,,\frac{x}{N}))=o(N)
$$

we conclude that

$$
|V_{N,k}(\eta)|\le C\sum_x|\bar\eta_{k,x}-\rho(t\,,\frac{x}{N})|^2+o(N)
$$

Now putting all the pieces together, if we can establish that for some $\delta>0$,

$$
\limsup_{k\to\infty}\limsup_{N\to\infty}\frac{1}{N}\log E^{\alpha_N(t)}\Big[\exp\big[\delta\sum_x|\bar\eta_{k,x}-\rho(t\,,\frac{x}{N})|^2\big]\Big]=0
$$

we would be done. This is done by large deviations. It is easy to estimate the above term by

$$\limsup_{k\to\infty}\frac{1}{k}\sup_{c\le\rho\le 1-c}\log E^{B_\rho}\big[\exp\delta\, k|\bar\eta_k-\rho|^2\big]=\sup_{c\le\rho\le 1-c}\sup_m[\delta|m-\rho|^2-h(m\,,\rho)]=0$$

for δ sufficiently small. Note that for some $a>0$,

$$h(m\,,\rho)\ge a(m-\rho)^2$$

It finally remains to establish local equilibrium.

Lemma 4.2 *Let $\mu_i(t)$ be the probabilities at time t for the states i of a finite Markov Chain with transition rate matrix $A=\{a_{i,j}\}$. Assume that α_i are the invariant probabilities and*

$$p_i(t)=\frac{\mu_i(t)}{\alpha_i}$$

are the densities. Then

$$2\int_0^\infty D(\sqrt{p(t)}\,)dt\le H(0).$$

Proof

$$\begin{aligned}\frac{dH(t)}{dt}&=\frac{d}{dt}\sum_i p_i(t)\log p_i(t)\alpha_i\\&=\sum_i[\log p_i(t)-1]p_i'(t)\alpha_i\\&=\sum_{i,j}p_j(t)\,a_{j,i}\log p_i(t)\alpha_j\\&=\sum_{i,j}\alpha_jp_j(t)\,a_{j,i}\log p_i(t)\\&=2\sum_{i,j}\alpha_jp_j(t)\,a_{j,i}\log\sqrt{p_i(t)}\\&\le 2\sum_j\alpha_j\sqrt{p_j(t)}\sum_i a_{j,i}\sqrt{p_i(t)}\\&=2<A\sqrt{p},\ \sqrt{p}>\\&=-2\mathcal{D}(\sqrt{p(t)}\,)\end{aligned}$$

Once we have control over the Dirichlet form, it does not matter what the dynamics is. The situation is the same as in the symmetric case with one big difference. There is a factor of N missing. The Dirichlet form in one dimension is bounded by C rather than CN^{-1}. This does not affect the local equilibrium principle as long we use the averages over long microscopic blocks. □

Remark Actually much stronger results are known in this special case. Instead of smooth solutions which exist only for short times, convergence to the physical (entropic) weak solution has been established for a wide class of models including the current one. However they all require some attractiveness assumptions due to the fact that coupling methods are used at some point. The relative entropy method works more or less universally, but needs both smooth solutions and special random initial conditions.

5 Nongradient systems

We now turn to the analysis of the asymmetric mean 0 case. Let us assume that we are in one dimension in order to simplify things a bit.

When we calculated

$$d[\frac{1}{N}\sum_x J(\frac{x}{N})\eta_t(x)]$$

we ended up with

$$\begin{aligned}&\frac{1}{N}\sum_x\sum_y \eta(x) < \frac{1}{2}[J'(\frac{x}{N}) + J'(\frac{y}{N})], N(1-\eta(y))(y-x)p(y-x) > dt + M_N(t)\\ &= [\frac{1}{2N}\sum_x J'(\frac{x}{N})N\Psi_x]dt + M_N(t)\end{aligned} \tag{5.1}$$

with $\Psi_x = W(\tau_x\eta)$ and

$$\begin{aligned}W(\eta) &= [\eta(0)\sum_z(1-\eta(z))z\,p(z) + (1-\eta(0))\sum_z \eta(-z)z\,p(z)]\\ &= [\eta(0)\sum_z(1-\eta(z))z\,p(z) - (1-\eta(0))\sum_z \eta(z)z\,p(-z)]\end{aligned} \tag{5.2}$$

Moreover $M_N(t)$ was negligible. However at this point we cannot do summation by parts and so we have a 'nongradient' model and the hydrodynamic limit can no longer be established by simple averaging. Some of the details of nongradient analysis can be found in [Q], where symmetric simple exclusion models with multiple types of particles are considered. The asymmetric mean zero case was considered in [L].

The analysis for the asymmetric mean zero case depends on an important fact that is stated and proved as Lemma 5.1. A n-cycle $C=\{0=x_0, x_1, x_2, \cdots, x_n = 0\}$ is a loop through $n-1$ distinct non zero integers that begins and ends at zero. With any such cycle we can associate a probability distribution π_C with mean 0:

$$\pi_C = \frac{1}{n}\sum_{i=1}^n \delta_{x_i - x_{i-1}}$$

While cycles of length 2 lead to symmetric distributions, in general, cycles of length 3 or more lead to asymmetric mean zero distributions. Since some of the increments $x_i - x_{i-1}$ could be repeated π_C is not necessarily the uniform distribution on some finite set.

Lemma 5.1 *Any probability distribution $p(\cdot)$ that is supported on a finite set of integers and has mean 0, can be written as a finite convex combination of distributions π_C arising from a collection of cycles.*

$$p = \sum_j \alpha_j \pi_{C_j}$$

Proof Let us denote the support of p by S. Suppose S consists of k points $a_1, \cdots, a_k$. We will show that there exist nonnegative integers $n_1, n_2, \cdots, n_k$ such that $\sum_j n_j a_j = 0$. In fact, because $d = 1$, it is obvious that we must have at least two a_j's say a_1 and a_2 that are of opposite signs. The choice of $n_1 = |a_2|, n_2 = |a_1|$

does it and gives us a cycle C of size $n_1 + n_2$ (n_1 steps of size a_1 and n_2 steps of size a_2 in any order). The support of π_C is contained in S. We can pick a weight α such that α is the largest possible positive number such that

$$\alpha\pi_C \leq p$$

Such a choice will allow us to write

$$p = \alpha\pi_C + (1-\alpha)p'$$

with p' being another mean zero probability distribution but supported on a set S' of cardinality at least one less than that of S. We can repeat the process till we are done. □

Given a cycle $C = \{0 = x_0, x_1, \cdots, x_n = 0\}$ of length n there is a special simple exclusion generator that corresponds to it

$$\mathcal{A}_C = \sum_x \mathcal{L}_{C+x}$$

that is the sum over translates of

$$(\mathcal{L}_C\, g)(\eta) = \frac{1}{n}\sum_{i=1}^{n} \eta(x_i)(1-\eta(x_{i+1}))[g(\eta^{x_i,x_{i+1}}) - g(\eta)] \tag{5.3}$$

For each C, $\mathcal{A}_C$ is the generator of a special asymmetric mean zero simple exclusion. For instance the cycle $\{0,1,2,0\}$ produces $p(1) = \frac{2}{3}$ and $p(-2) = \frac{1}{3}$. Finally the lemma allows us to write for any p a representation

$$\mathcal{A} = \sum_j \alpha_j \mathcal{A}_{C_j} = \sum_{j,x} \alpha_j \mathcal{L}_{C_j+x} \tag{5.4}$$

The important point here is each $\mathcal{L}_{C_j+x}$, while slightly more complicated than simple exchanges $[g(\eta^{x,y}) - g(\eta)]$ across a bond, are still finite dimensional objects and our invariant measures are invariant under these.

For each k the uniform distribution ν_k on the $\binom{n}{k}$ subsets of cardinality k is an invariant measure for both $\mathcal{L}_C$ and $\mathcal{L}_{C,sym}$. The Dirichlet form is given by

$$D_{C,k}(f) = \int \frac{1}{4n}\sum_i [f(\eta^{x_{i-1},x_i}) - f(\eta)]^2 d\nu_k$$

or more generally for any invariant measure μ

$$D_{C,\mu}(f) = \int \frac{1}{4n}\sum_i [f(\eta^{x_{i-1},x_i}) - f(\eta)]^2 d\mu \tag{5.5}$$

Since the set of invariant measures for $\mathcal{L}_{C,sym}$ and $\mathcal{L}_C$ are the same, by linear algebra, the equation

$$\mathcal{L}_{C,sym} f_C = \mathcal{L}_C f \tag{5.6}$$

can be solved and we can have estimates of the form

$$D_{C,k}(f_C) \leq C D_{C,k}(f) \tag{5.7}$$

$$\|f_C\|_\infty \leq C\|\mathcal{L}_C f\|_\infty \tag{5.8}$$

with a constant that depends only on n.

We can use the representation (5.4) and the existence of solutions to (5.6) with estimates (5.7) and (5.8) to estimate various objects. For instance we have,

Theorem 5.2 *There exists a constant C depending only on $p(\cdot)$ such that*

$$|\int (\mathcal{A}f)(\eta)g(\eta)d\mu(\eta)| \leq C\sqrt{D_\mu(f)}\sqrt{D_\mu(g)}.$$

Proof We write $\mathcal{A}$ as the sum

$$\mathcal{A} = \sum_{j,x} \alpha_j \mathcal{L}_{C_j+x}$$

and try to estimate

$$\int (\mathcal{L}_{C_j,x} f)(\eta)g(\eta)d\mu(\eta)$$

by first solving

$$\mathcal{L}_{C_j+x,sym}\ f_{j,x} = \mathcal{L}_{C_j+x} f$$

and getting good estimates on $f_{C_j,x}$ of the form (5.7). Then

$$\begin{aligned}
|\int (\mathcal{A}f)(\eta)g(\eta)d\mu(\eta)| &= \sum_{x,j} \alpha_j |\int (\mathcal{L}_{C_j+x} f)(\eta)g(\eta)d\mu(\eta)| \\
&= \sum_{x,j} \alpha_j |\int (\mathcal{L}_{C_j+x,sym} f_{j,x})(\eta)g(\eta)d\mu(\eta)| \\
&\leq \sum_{j,x} \alpha_j \sqrt{D_{j,x}(f_{j,x})}\sqrt{D_{j,x}(g)} \\
&\leq \sqrt{\sum_{j,x} \alpha_j D_{j,x}(f_{j,x})}\sqrt{\sum_{j,x} \alpha_j D_{j,x}(g)} \\
&\leq C\sqrt{\sum_{j,x} \alpha_j D_{j,x}(f)}\sqrt{\sum_{j,x} \alpha_j D_{j,x}(g)} \\
&\leq C\sqrt{D(f)}\sqrt{D(g)}
\end{aligned}$$

□

Remark This estimate is valid in the infinite volume for any Bernoulli measure μ_ρ with a constant C independent of ρ or in the periodic case $\mathbf{Z}_N$ for the uniform distribution of k particles, with a constant independent of N and k.

Remark The Dirichlet form $D(f)$ depends on $p(\cdot)$. But as long as the Random Walk induced by the symmetrized version of p is irreducible, they are all equivalent.

Another use of the representation is an estimate on the Feynaman-Kac eigenvalue. But first we need an elementary Lemma.

Lemma 5.3 *Let P_μ be a Markov Process with generator $\mathcal{A}$ which is in equlibrium having for its marginals an inavriant measure μ.*

$$E^{P_\mu}\Big[\exp\big[\int_0^t f(\eta(s))ds\big]\Big] \leq \exp[t\,\sigma(f)]$$

where

$$\sigma(f) = \sup_{\|g\|_{2,\mu}=1} \Big[E^\mu[f(\eta)g^2(\eta)] - D(g)\Big]$$

$$= \sup_{\lambda \in \mathcal{M}} \left[\sum f(\eta)\lambda(\eta) - I(\lambda) \right]$$

where the supremum is taken over all probability distributions λ and $I(\lambda)$ is defined as

$$I(\lambda) = D\left(\sqrt{\frac{d\lambda}{d\mu}}\right) = -\langle \frac{\mathcal{A}+\mathcal{A}^*}{2} \sqrt{\frac{d\lambda}{d\mu}}\, , \sqrt{\frac{d\lambda}{d\mu}} \rangle_\mu$$

Remark Note that although the expectation involves a process with generator $\mathcal{A}$ which may not be symmetric with respect to μ, the variational formula involves only the symmetric part of $\mathcal{A}$.

Proof If we define

$$u(t\,,x) = E_x\left[\exp\left[\int_o^t f(x(s))ds \right] \right]$$

then $u(t\,,x)$ solves

$$\frac{\partial u}{\partial t} = \mathcal{A}u + fu; \quad u(0\,,x) = 1$$

and therfore

$$\begin{aligned} \frac{d}{dt}\int u^2 d\mu &= \int 2u\frac{du}{dt}d\mu = \int 2u[\mathcal{A}u + fu]d\mu \\ &= 2[\int fu^2 d\mu - D(u)] \\ &\le 2\|u\|_{2,\mu}^2 \sigma(f) \end{aligned}$$

providing us the estimate

$$\|u(t\,,\cdot)\|_{2,\mu} \le \exp[t\,\sigma(f)]$$

Since

$$E^\mu\left[\exp\left[\int_o^t f(x(s))ds \right] \right] = \int u(t\,,x)d\mu \le \|u(t\,,\cdot)\|_{2,\mu}$$

we are done. □

Theorem 5.4 *Let f be arbitrary and $g = \mathcal{A}f$. Then*

$$\sigma(g) = \sigma(\mathcal{A}f) = \sup_{\|\phi\|_{2,\mu}=1} \left[\int g\phi^2 d\mu - D(\phi) \right]$$

satisfies the bound

$$\sigma(\mathcal{A}f) \le C \sum_{x,j} \|\mathcal{L}_{x,j} f\|_\infty^2$$

Proof For arbitrary ϕ with $\|\phi\|_{2,\mu} = 1$ we need to estimate

$$\begin{aligned}
&\int (\mathcal{A}f)\phi^2 d\mu - D(\phi) \\
&= \sum_{x,j} \alpha_j \Big[\int (\mathcal{L}_{x,j} f)\phi^2 d\mu - D_{x,j}(\phi) \Big] \\
&= \sum_{x,j} \alpha_j \Big[\int (\mathcal{L}_{x,j,sym} f_{x,j})\phi^2 d\mu - D_{x,j}(\phi) \Big] \\
&= \sum_{x,j} \alpha_j \Big[\int [\sum_{b \in C_{x,j}} \frac{1}{2n_j} [f_{x,j}(\eta^b) - f_{x,j}(\eta)]] \phi^2 d\mu - D_{x,j}(\phi) \Big] \\
&= -\frac{1}{2} \sum_{x,j} \alpha_j \Big[\int [\sum_{b \in C_{x,j}} \frac{1}{2n_j} [f_{x,j}(\eta^b) - f_{x,j}(\eta)]] [\phi^2(\eta^b) - \phi^2(\eta)] d\mu - D_{x,j}(\phi) \Big] \\
&= -\frac{1}{2} \sum_{x,j} \alpha_j \Big[\int [\sum_{b \in C_{x,j}} \frac{1}{2n_j} [f_{x,j}(\eta^b) - f_{x,j}(\eta)]] [\phi(\eta^b) - \phi(\eta)][\phi(\eta^b) + \phi(\eta)] d\mu \\
&\qquad\qquad - D_{x,j}(\phi) \Big] \\
&\leq C \sum_{x,j} \|\mathcal{L}_{x,j} f\|_\infty^2
\end{aligned}$$

We have represented $\mathcal{L}_{x,j} f$ as $\mathcal{L}_{x,j,sym} f_{x,j}$ and estimated $\|f_{x,j}\|_\infty$ interms of $\|\mathcal{L}_{x,j} f\|_\infty$ by using (5.8). □

Now we return to the analysis of (5.1). It would be nice if W were a gradient, perhaps even of the form $a\,(\eta(0) - \eta(1))$ for some constant $a > 0$. A summation by parts will give us the heat equation. At least we can hope that $W - a\,(\eta(0) - \eta(1))$ is negligible in some sense and so can be thrown away. It is indeed true that there is such a constant a, but it depends on the local density ρ and is really a function $a(\rho)$. The final equation then is

$$\frac{\partial \rho(t,\theta)}{\partial t} = \frac{1}{2} \frac{\partial}{\partial \theta} a(\rho(t,\theta)) \frac{\partial \rho(t,\theta)}{\partial \theta} \tag{5.9}$$

The first question is the sense in which $W - a(\eta(0) - \eta(1))$ is negligible. This will be formulated by means of a property that will hold relative to the processes $\{P_\rho\}$ in equilibria with marginals $\{\mu_\rho\}$ i.e. the Bernoulli meaasure with density ρ on $[0,1]^{\mathbf{Z}}$. Let us consider the class $\mathcal{F}$ of local functions $f(\eta)$ that have mean zero under every μ_ρ. If we denote by $\mu_{N,n}$, the uniform distribution on all configurations on N sites that have n occupied sites, for any local function f measurable relative to these sites, we have

$$E^\rho[f] = \sum \binom{N}{n} \rho^n (1-\rho)^{N-n} E^{\mu_{N,n}}[f]$$

Therefore $E^\rho[f] \equiv 0$ in ρ if and only if $E^{\mu_{N,n}}[f] \equiv 0$ for all $0 \leq n \leq N$ and N large enough for f to be measurable. Examples of such functions are W, $\eta(0) - \eta(1)$ and a whole family $f = \mathcal{A}g$ where $\mathcal{A}$ is the generator and g is any local function. Suppose $W = \mathcal{A}g + a(\eta(0) - \eta(1))$ for some local g. The only problem in replacing

W by $a(\eta(0)-\eta(1))$ is the disposition of the difference

$$\int_0^t \mathcal{A}\Big[\sum_x J'\big(\frac{x}{N}\big)g(\tau_x\eta_s)\Big]\,ds \tag{5.10}$$

We apply Ito's formula to the function

$$\frac{1}{N^2}\sum_x J\big(\frac{x}{N}\big)g(\tau_x\eta)$$

and because the generator has an extra factor of N^2 due to the speeded up time scale we get

$$\begin{aligned}\frac{1}{N^2}\sum_x J'\big(\frac{x}{N}\big)g(\tau_x\eta_t) &- \frac{1}{N^2}\sum_x J'\big(\frac{x}{N}\big)g(\tau_x\eta_0)\\ &= \int_0^t \mathcal{A}\Big[\sum_x J'\big(\frac{x}{N}\big)g(\tau_x\eta_t)\Big]\,dt + M_N(t)\end{aligned} \tag{5.11}$$

An elementary computation shows that the martingale term is negligible for large N. So we see that the term (5.10) can be safely ignored. What we really need is a theorem to the effect that there is a function $a(\rho)$ and local functions $g_\rho(\eta)$ such that it is almost true that $W = \mathcal{A}g_\rho + a(\rho)(\eta(0)-\eta(1))$ relative to P_ρ (what ever that means!)

Let $f \in \mathcal{F}$ be measurable with respect to sites on $[-\ell,\ell]$. We can consider the central limit theorem for

$$\frac{1}{\sqrt{t}}\int_0^t f(\eta_s)\,ds$$

under P_ρ or some $P_{\mu_{N,n}}$. We shall assume that the dynamics is some reversible version of our processs, perhaps even just the nearest neighbor symmetric simple exclusion process. We will have an asymptotic variance that we denote by $\sigma^2(N,n,f)$. We can replace $f(\eta)$ by $f_N(\eta) = \sum_{|i|\le N-\ell}^n f(\tau_i\,\eta)$ where ℓ is chosen so that f is measurable on $[-\ell,\ell]$. The corresponding variance will be denoted by $\sigma_N^2(n\,,f)$. The main step in the analysis of nongradient systems is the following Theorem 5.5 which will be proved in the next lecture.

Theorem 5.5 *Let $\mathcal{G}$ denote the set of functions $g = g(\rho\,,\eta) = g_\rho(\eta)$ that are local and depend smoothly on ρ. There exists a continuous function $a(\rho)$ such that*

$$\inf_{g\in\mathcal{G}}\limsup_{N\to\infty}\sup_{0\le n\le 2N+1}\frac{\sigma_N^2\big(n\,,W - a(\frac{n}{2N+1})[\eta(0)-\eta(1)] - \mathcal{A}g_{\frac{n}{2N+1}}\big)}{2N+1} = 0$$

Remark Note that because $N\to\infty$, this is a theorem concerning the behavior under various stationary equilibrium processes $\{P_\rho\}$ although we need to consider nonequilibrium situations while trying to establish hydrodynamic limits.

We will proceed with our analysis and see why Theorem 5.5 is sufficient.

Let P_N be the process on the time interval $[0,T]$ starting from some configuration η_0. We define

$$\xi_{N,\epsilon,\delta} = \int_0^T \sum_x J'(\frac{x}{N})W(\tau_x\eta_t)dt - \int_0^T \mathcal{A}\big[\sum_x J'(\frac{x}{N})g(\bar{\eta}_{N\epsilon,x,t}\,,\tau_x\eta_t)\big]\,dt$$
$$-\int_0^T \frac{1}{2N\delta}\sum_x J'(\frac{x}{N})a(\bar{\eta}_{N\epsilon,x,t})[\eta_t(x-N\delta)-\eta_t(x+N\delta)]dt$$

It is clearly sufficient to show that

$$\inf_{g\in\mathcal{G}}\limsup_{\epsilon\to 0}\limsup_{\delta\to 0}\limsup_{N\to\infty} E^{P_N}\big[|\xi_{N,\epsilon,\delta}|\big]=0 \tag{5.12}$$

It is not difficult to see that the term

$$\int_0^T \mathcal{A}\big[\sum_x J(\frac{x}{N})g(\bar{\eta}_{N\epsilon,x,t}\,,\tau_x\eta_t)\big]\,dt$$

continues to be negligible and therefore

$$\frac{1}{2}\int_0^T \sum_x J'(\frac{x}{N})W(\tau_x\eta_t)dt$$

can be replaced by

$$\lim_{\epsilon\to 0}\lim_{\delta\to 0}\frac{1}{2}\int_0^T \frac{1}{2N\delta}\sum_x J'(\frac{x}{N})a(\bar{\eta}_{N\epsilon,x,t})[\eta_t(x-N\delta)-\eta_t(x+N\delta)]dt$$
$$=\lim_{\epsilon\to 0}\lim_{\delta\to 0}\frac{1}{2}\int_0^T\int_{\mathbf{T}}\frac{1}{2\delta}J'(\theta)a\Big(\frac{1}{2\epsilon}\int_{\theta-\epsilon}^{\theta+\epsilon}\rho(t\,,w)dw\Big)[\rho(t\,,\theta-\delta)-\rho(t\,,\theta+\delta)]d\theta dt$$
$$=\lim_{\epsilon\to 0}-\frac{1}{2}\int_0^T\int_{\mathbf{T}}J'(\theta)a\Big(\frac{1}{2\epsilon}\int_{\theta-\epsilon}^{\theta+\epsilon}\rho(t\,,w)dw\Big)\rho'(t\,,\theta)d\theta dt$$
$$=-\frac{1}{2}\int_0^T\int_{\mathbf{T}}J'(\theta)a(\rho(t\,,\theta))\rho'(t\,,\theta)d\theta dt$$

giving us a weak formulation of (5.9).

An actual proof will proceed in the following manner. We consider the family of processes P_N on $\Omega_N = D[[0,T];[0,1]^{\mathbf{Z}_N}]$ with generator $N^2\mathcal{A}$ and an arbitrary initial configuration $\eta_o(\cdot)$. We view the process of empirical distributions

$$\nu_N(t)=\frac{1}{N}\sum_x \delta_{\frac{x}{N}}\eta_t(x) \qquad 0\le t\le T$$

as a stochastic process with values in $\mathcal{M}(\mathbf{T})$, the space of measures on $\mathbf{T}$ and call the induced distribution $\widehat{P}_N$. We prove that the family $\widehat{P}_N$ is tight and if $\widehat{P}$ is any limit point it is supported on measure valued maps $\nu(t)$ that are weakly continuous in t and have densities $\rho(t,\theta)$ satisfying $0\le\rho(\cdot,\cdot)\le 1$. Moreover

$$E^P\Big[\int_0^T\int_{\mathbf{T}}|\rho_\theta(t,\theta)|^2 dt d\theta\Big]<\infty \tag{5.13}$$

Once we have (5.12) and the estimate (5.13) we see that any limit point $\widehat{P}$ is concentrated on the set of weak solutions of (5.9) satisfying the energy estimate

$$\int_0^T \int_{\mathbf{T}} |\rho_\theta(t,\theta)|^2 dt d\theta < \infty \tag{5.14}$$

Moreover if the initial condition $\eta_0 = \eta_0^N(\cdot)$ is such that

$$\lim_{N\to\infty} \frac{1}{N} \sum_x J(\frac{x}{N})\eta_0^N(x) = \int_{\mathbf{T}} J(\theta)\rho_0(\theta)d\theta$$

then $\widehat{P}$ is supported by weak solutions with initial condition $\rho_0(\cdot)$. Such solutions are unique according to Lemma 5.6. and the hydrodynamic limit is established because we see that $\widehat{P}$ is supported on the single unique solution.

Therefore it all comes down to proving tightness, (5.12) and (5.13). First we will prove the tightness of the family $\widehat{P}_N$. Since the martingale part is small in the mean square sense, by Doob's inequlaity it is small uniformly in t for every starting configuration. Tightness will follow from superexponential estimates on the modulus of continuity of

$$\zeta_J(t) = \int_0^t \sum_x J(\frac{x}{N})\Psi_x(\eta(s))ds$$

in equlibrium. Clearly an estimate of the form

$$E^N\Big[\exp\big[N\int_0^t \sum_x J(\frac{x}{N})\Psi_x(\eta(s))ds\big]\Big] \le \exp[NC_J t]$$

on the equilibrium process will do the trick. This in turn can be derived from the variational estimate

$$\sup_{\|\phi\|_{\mu_N}=1} \Big[E^{\mu_N}\big[\phi^2 \sum_x NJ(\frac{x}{N})\Psi_x(\eta)\big] - N^2 D_N(\phi)\Big] \le NC_J t$$

using theorem 5.4. Basically $f = \sum_y y\eta(y)$. The same method provides the estimate

$$\sup_{\|\phi\|_{\mu_N}=1} \Big[E^{\mu_N}\big[\phi^2 \sum_x NJ(\frac{x}{N})[\eta(x+1)-\eta(x)]\big] - N^2 D_N(\phi)\Big] \le NC_J t$$

The constant C_J is essentially $C\int J^2(\theta)d\theta$. Since, for any initial distribution, the entropy relative to the equilibrium μ_N is atmost N, the standard entropy inequality implies that

$$E^{\widehat{P}_N}\Big[\int_0^T \frac{1}{N}\sum_x J'(\frac{x}{N})\eta(x,s)ds\Big] \le C + \frac{T}{2}\int J^2(\theta)d\theta$$

In fact one can do the time dependent potentials as well and get

$$E^{\widehat{P}_N}\Big[\int_0^T \frac{1}{N}\sum_x J'(t,\frac{x}{N})\eta(x,s)ds\Big] \le C + \frac{1}{2}\int\int_0^T J^2(t,\theta)d\theta$$

and (5.13) folows not so routinely from this estimate. If we are going to use entropy inequality, then

$$\begin{aligned} E^Q[\max(X,Y)] &\le \log E^P[\max(e^X, e^Y)] + h(Q,P) \\ &\le \log E^P[e^X + e^Y] + h(Q,P) \\ &\le \log 2 + \max\{\log E^P[e^X], \log E^P[e^Y]\} + h(Q,P) \end{aligned}$$

and if we are going to divide by N and pass to the limit $\log 2$ disappears from sight. It is almost as though the supremum can be moved inside the integral freely. So for the limit instead of just

$$\sup_J \left[E^{\widehat{P}} \left[\int_0^T \int J'(t,\theta)\rho(\theta\,,s) ds d\theta - \frac{1}{2} \int \int_0^T J^2(t,\theta) d\theta \right]\right] \le C$$

we actually can get

$$E^{\widehat{P}} \left[\sup_J \left[\int_0^T \int J'(t,\theta)\rho(\theta\,,s) ds d\theta - \frac{1}{2} \int \int_0^T J^2(t,\theta) d\theta \right]\right] \le C$$

which is just (5.13)

We will prove (5.12) by showing

$$\inf_{g\in\mathcal{G}} \limsup_{\epsilon\to 0} \limsup_{\delta\to 0} \limsup_{N\to\infty} \frac{1}{N} \log E^{Q_N}\left[\exp\left[\lambda\, N\, \xi_{N,\epsilon,\delta}\right]\right] = 0 \tag{5.15}$$

for every real λ. Since we can replace J by λJ there is no loss of generality in taking $\lambda = 1$. We can as well replace J' by an arbitrary J and consider

$$\xi_{N,\epsilon,\delta} = \int_0^T \sum_x J(\frac{x}{N}) G_{N,\epsilon,\delta}(\tau_x \eta_t) dt$$

with

$$G_{N,\epsilon,\delta}(\eta) = W(\eta) - \mathcal{A}g(\bar{\eta}_{N\epsilon,0}\,, \eta) - \frac{1}{2N\delta} a(\bar{\eta}_{N\epsilon,0})[\eta(-N\delta) - \eta(N\delta)]$$

Since Q_N is starting from equlibrium we have the following estimates for the Feynman-Kac integrals by the variational formula. See Lemma 5.3.

$$\begin{aligned} \frac{1}{N} &\log E^{Q_N}\left[\exp\left[N\, \xi_{N,\epsilon,\delta}\right]\right] \\ &\le \frac{T}{N} \sup_{\|\phi\|_{2,\mu}=1} \left[E^\mu[\phi^2(\eta) \sum_x N\, J(\frac{x}{N}) G_{N,\epsilon,\delta}(\tau_x\eta)] - N^2\, \mathcal{D}_N(\phi)\right] \\ &= T \sup_{\|\phi\|_{2,\mu}=1} \left[E^\mu[\phi^2(\eta) \sum_x J(\frac{x}{N}) G_{N,\epsilon,\delta}(\tau_x\eta)] - N\mathcal{D}_N(\phi)\right] \end{aligned} \tag{5.16}$$

Basically we need to show that

$$\inf_{g\in\mathcal{G}} \limsup_{\epsilon\to 0} \limsup_{\delta\to 0} \limsup_{N\to\infty} \sup_{\|\phi\|_{2,\mu}=1} \left[E^\mu[\phi^2(\eta) \sum_x J(\frac{x}{N}) G_{N,\epsilon,\delta}(\tau_x\eta)] - N\mathcal{D}_N(\phi)\right] = 0 \tag{5.17}$$

This is a matter of controlling the first term by the second term. The first term and the Dirichlet form may be split into pieces, with control over individual pieces. We also note that the Dirichlet form $\mathcal{D}_N(\phi)$ naturally splits as the sum $\sum_x \mathcal{D}_{N,x,x+1}(\phi)$ of contributions over the individual bonds. We can pretend that our Dirichlet form is the nearest neighbor one because they are in any case equivalent. Since J is arbitrary and can be replaced by a constant multiple of itself, for obtaining estimates, it does not matter if we replace the Dirichlet form by an equivalent one. We replace $G_{N,\epsilon,\delta}$ by $H_{N,k}$ where

$$H_{N,k}(\eta) = \frac{1}{2k+1}\sum_{|x|\le k-\ell} W(\tau_x\eta) - \frac{1}{2k+1}\sum_{|x|\le k-\ell} \mathcal{A}g(\bar\eta_{k,0},\tau_x\eta) - \frac{1}{2k+1}a(\bar\eta_{k,0})[\eta(-k)-\eta(k)]$$

Here ℓ is fixed integer depending on the local function W and g and is picked so that the function $H_{N,k}(\eta)$ depends only the $\{\eta(x)\}$ with $|x|\le k$ For the difference $F_{N,k,\epsilon,\delta} = G_{N,\epsilon,\delta} - H_{N,k}$ we show that for every $\beta>0$,

$$\limsup_{\epsilon\to 0}\limsup_{\delta\to 0}\limsup_{k\to\infty}\limsup_{N\to\infty}\sup_{\|\phi\|_{2,\mu}=1}\Big[E^{\mu}\big[\phi^2(\eta)\sum_x J(\frac{x}{N})F_{N,\epsilon,\delta,k}(\tau_x\eta)\big] - \beta N\mathcal{D}_N(\phi)\Big] = 0 \tag{5.18}$$

To see this we rewrite $F_{N,k.\epsilon,\delta}$ as the sum of several terms and control each one of them in terms of the Dirichlet form. The first one takes the form

$$\sum_x \Big[\frac{1}{2k+1}\sum_{|y-x|\le k}[J(\frac{y}{N}) - J(\frac{x}{N})]\Big]W(\tau_x\eta)$$

that is easily controlled. The second term is

$$\sum_x J(\frac{x}{N})\big[\frac{1}{2N\delta}a(\bar\eta_{N\epsilon,x})[\eta(x-N\delta)-\eta(x+N\delta)] - \frac{1}{2k}a(\bar\eta_{k,x})[\eta(x-k)-\eta(x+k)]\big]$$

that can be rewritten as

$$\sum_x A_N(\epsilon,\delta,k,x)[\eta(x-1)-\eta(x)]$$

where

$$A_N(\epsilon,\delta,k,x) = \frac{1}{2N\delta}\sum_{x-N\delta+1\le y\le x+N\delta} J(\frac{y}{N})a(\bar\eta_{N\epsilon,y}) - \frac{1}{2k+1}\sum_{x-k+1\le y\le x+k} J(\frac{y}{N})a(\bar\eta_{k,y}).$$

For any $\beta>0$ there exists a constant C_β such that

$$\sup_{\|\phi\|_{2,\mu_N}=1} E^{\mu_N}\Big[A_N(\epsilon,\delta,k,x)\phi^2(\eta)[\eta(x+1)-\eta(x)] - \beta N\mathcal{D}_{N,x,x+1}(\phi) - \frac{C_\beta}{N}[A_N(\epsilon,\delta,k,x)]^2\Big] \le 0 \tag{5.19}$$

and therefore with the help of (5.19), we can reduce (5.18) to estimating

$$\sup_{\|\phi\|_{2,\mu_N}=1}\left[E^{\mu_N}\left[\phi^2(\eta)\frac{1}{N}\sum_x [A_N(\epsilon,\delta,k,x)]^2\right]-\beta N\mathcal{D}_N(\phi)\right]$$

This can be done using the continuity of $a(\cdot)$ and $J(\cdot)$ and the two block estimate. The last term is the difference

$$S_{\epsilon,k}(\eta)=\sum_x J(\frac{x}{N})\Big[\frac{1}{2k+1}\sum_{y:|y-x|\le k-\ell}\big[\mathcal{A}g(\bar\eta_{k,y},\tau_y\eta)-\mathcal{A}g(\bar\eta_{N\epsilon,x},\tau_x\eta)\big]\Big]$$

An elmentary calculation shows that

$$\|\mathcal{L}_{x,j}S_{\epsilon,k}\|_\infty\le C\left[\frac{1}{N\epsilon}+\frac{1}{k}\right]$$

and we can use Theorem 5.4 to prove

$$\limsup_{\epsilon\to0}\limsup_{k\to\infty}\limsup_{N\to\infty}\sup_{\|\phi\|=1}\left[E^{\mu}\left[\phi^2(\eta)S_{\epsilon,k}(\eta)\right]-\beta N\mathcal{D}_N(\phi)\right]=0 \tag{5.20}$$

In order to complete the argument we will have to see why Theorem 5.5 implies that

$$\inf_{g\in\mathcal{G}}\limsup_{k\to\infty}\limsup_{N\to\infty}\sup_{\|\phi\|_{2,\mu_N}=1}\left[E^{\mu}_N\left[\phi^2(\eta)\sum_x J(\frac{x}{N})H_{N,k}(\tau_x\eta)\right]-N\mathcal{D}_N(\phi)\right]=0 \tag{5.21}$$

We can write

$$D_N(\phi)=\frac{1}{2k}\sum_x D_{x,k}(\phi)=\frac{1}{2k}\sum_x\sum_{b\in(x-k,x+k)}D_b(\phi)$$

and

$$\begin{aligned}
&\sup_{\|\phi\|_{2,\mu_N}=1}\left[E^{\mu_N}\left[\phi^2(\eta)\sum_x J(\frac{x}{N})H_{N,k}(\tau_x\eta)\right]-N\mathcal{D}_N(\phi)\right]\\
&=\frac{1}{2k+1}\sup_{\|\phi\|_{2,\mu_N}=1}\left[E^{\mu_N}\left[\phi^2(\eta)\sum_x J(\frac{x}{N})(2k+1)H_{N,k}(\tau_x\eta)\right]\right.\\
&\qquad\qquad\left.-N(2k+1)\mathcal{D}_N(\phi)\right]\\
&\le\frac{1}{2k+1}\sum_x\sup_{\|\phi\|_{2,\mu_N}=1}\left[E^{\mu_N}\left[\phi^2(\eta)J(\frac{x}{N})(2k+1)H_{N,k}(\tau_x\eta)\right]-N\mathcal{D}_{x,k}(\phi)\right]\\
&\le\frac{N}{2k+1}\sum_x\sup_{\|\phi\|_{2,\mu_N}=1}\left[E^{\mu_N}\left[\phi^2(\eta)\frac{1}{N}J(\frac{x}{N})(2k+1)H_{N,k}(\tau_x\eta)\right]-\mathcal{D}_{x,\ell}(\phi)\right]
\end{aligned}$$

Taking limits as $N\to\infty$ and using the perturbation expansion of the eigenvalue in the small parameter $\frac{1}{N}$, we estimate the limsup of left hand side by

$$C_J\sup_n\frac{1}{2k+1}\sigma_k^2(n,(2k+1)H_{N,k}(\eta))$$

where

$$C_J=\int J^2(\theta)d\theta$$

and $\sigma_k^2(n\,,(2k+1)H_{N,k}(\eta))$ is the variance of

$$\sum_{|x|\le k-\ell} W(\tau_x\eta) - \sum_{|x|\le k-\ell} \mathcal{A}g(\bar\eta_{k,0},\tau_x\eta) - a(\bar\eta_{k,0})[\eta(-k)-\eta(k)]$$

calculated for the uniform distribution on the configurations in a block of size $2k+1$ with n particles. If $k\to\infty$ and $\frac{n}{2k+1}\to\rho$ we are back in the situation of Theorem 5.5

Remark We have the genrator $\mathcal{A}$ and its symmetric version $\mathcal{A}_{sym}$ that share the same Dirichlet form. The variational formula, actually gives the eigenvalues for $\mathcal{A}_{sym}$ and perturbatively for large N they are controlled by the variance in the central limit theorem for the symmetric process. It is a fact for the mean zero case, that if we write $\mathcal{A}_N=\mathcal{A}_{sym,N}+\mathcal{A}_{skew,N}$, then from theorem 5.2

$$\|\mathcal{A}_{skew,N}\|_{H^1\to H^{-1}}\le C$$

uniformly in N, and this in turn means a uniform bound of the form

$$\sigma^2(f)\le\sigma_{sym}^2(f)\le C\sigma^2(f)$$

Therefore while we have to use $\mathcal{A}$ for the exact eavluation of $a(\rho)$, for the purpose of estimation we can use $\mathcal{A}_{symm}$ and we will not lose anything in the process.

6 Nongradient case (continuation)

It only remains to prove theorem 5.5.

Theorem 5.5 *Let $\mathcal{G}$ denote the set of functions $g=g(\rho\,,\eta)=g_\rho(\eta)$ that are local and depend smoothly on ρ. There exists a function $a(\rho)$ such that*

$$\inf_{g\in\mathcal{G}}\limsup_{N\to\infty}\sup_{0\le n\le 2N+1}\frac{\sigma_N^2(n\,,W-a(\frac{n}{2N+1})[\eta(0)-\eta(1)]-\mathcal{A}g_{\frac{n}{2N+1}})}{2N+1}=0.$$

σ_N^2 can be taken relative to any reasonable symmetric simple exclusion process and we will for simplicity pick the nearest neighbor one. The big reduction now is that we have only to deal asymptotically, as $N\to\infty$, with different equilibria having n particles among $2N+1$ sites.

We will prove a stronger version, namely

Theorem 6.1 *For each $f\in\mathcal{F}$ the limit*

$$\lim_{\substack{N\to\infty\\ \frac{n}{2N+1}\to\rho}}\frac{\sigma_N^2(n,f)}{2N+1}=\langle\langle f,f\rangle\rangle_\rho$$

exists and for each f there exists $a(f,\rho)$ such that

$$\inf_{g\subset\mathcal{G}}\langle\langle f-a(f,\rho)(\eta(0)-\eta(1))-\mathcal{A}g,f-a(f,\rho)(\eta(0)-\eta(1))-\mathcal{A}g\rangle\rangle_\rho=0$$

In particular $a(\rho)=a(W,\rho)$. Moreover if we take $g=g(\rho,\eta)$ from $\mathcal{G}_s$, the class of smooth local functions, we have

$$\inf_{g\in\mathcal{G}_s}\sup_\rho\langle\langle f-a(f,\rho)(\eta(0)-\eta(1))-\mathcal{A}g,f-a(f,\rho)(\eta(0)-\eta(1))-\mathcal{A}g\rangle\rangle_\rho=0.$$

There is a basic formalism which is important. Let us fix ρ and the Bernoulli measure μ_ρ which is an invariant measure for the simple exclusion dynamics. Given a local function $f(\eta)$ we consider the formal sum

$$F(\eta) = \sum_x f(\tau_x \eta).$$

F does not really exist. But, for any two sites a, b

$$\xi_f^{a,b}(\eta) = F(\eta^{a,b}) - F(\eta)$$

makes perfectly good sense. Moreover, they are covariant, in the sense that

$$\xi_f^{x+a,x+b}(\eta) = \xi_f^{a,b}(\tau_x \eta)$$

and in fact the entire collection $\xi_f^{a,b}$ can be determined algebraically from $\xi_f^{0,1}(\eta)$ which itself must satisfy some simple algebraic conditions. We can look at the space of such abstract objects $\{\xi^{0,1}\}$ as a closed subspace $\mathcal{H} \subset L_2(\mu_\rho)$.

In addition to the ξ_f generated by local functions f, there is an additional object in $\mathcal{H}$ that is important.

$$\xi^{a,b} = (b-a)[\eta(a) - \eta(b)]$$

this comes formally from the nonlocal nonexistent f

$$f(\eta) = \sum_{x \le 0} \eta(x)$$

or equally formally, with

$$F(\eta) = \sum_x f(\tau_x \eta) = \sum_x x\,\eta(x)$$

Although $F(\tau\eta) \ne F(\eta)$ we do have $F(\tau\eta) - F(\eta) = -\sum_x \eta(x)$ which is formally conserved. The important fact is that $F(\eta^{0,1}) - F(\eta) = \xi_f^{0,1} = \eta(0) - \eta(1) \in \mathcal{H}$.

Proof of Theorem 6.1 Given $f \in \mathcal{F}$ which is a local function on $[-k, k]$ we solve in an interval $[-N, N]$ the equation

$$(\mathcal{A}_N^0 v_N)(\eta) = \frac{1}{2} \sum_{(x,x+1)\in[-N,N]} [v_N(\eta^{x,x+1}) - v_N(\eta] = \sum_{|x| \le N-k} f(\tau_x \eta)$$

where $\mathcal{A}_N^0$ is the restriction of the nearest neighbor symmetric simple exclusion process to the interval $[-N, N]$. The set of configurations with a given number of particles are the invariant subsets, with uniform distributions over all such classes of configurations being the invariant measures.

$$\begin{aligned}
\sigma_N^2(n, f) &= -2 < (\mathcal{A}_N^0)^{-1} \sum_x f(\tau_x \eta) \,, \sum_x f(\tau_x \eta) >_{N,n} \\
&= -2 < v_N \,, \mathcal{A}_N^0 v_N >_{N,n} \\
&= 2 D_{N,n}(v_N) \\
&= E^{\nu_{N,n}} \Big[\sum_{(x,x+1)\in[-N,N]} [v_N(\eta^{x,x+1}) - v_N(\eta)]^2 \Big]
\end{aligned}$$

Let us try some examples. Suppose $f = \mathcal{A}^0 g$ for some local $g \in \mathcal{G}$. Then

$$\mathcal{A}_N^0 \sum_x \tau_x g = \sum_x \tau_x f$$

and

$$\lim_{\substack{N \to \infty \\ \frac{n}{2N+1} \to \rho}} \frac{1}{2N} \sigma_N^2(n, f) = E^{\mu_\rho}[\xi_g^2]$$

or in polarized form if $f_i = \mathcal{A}^0 g_i$ for $i = 1, 2$

$$\lim_{\substack{N \to \infty \\ \frac{n}{2N+1} \to \rho}} \frac{1}{2N} Cov_N(n, f_1, f_2) = E^{\mu_\rho}[\xi_{g_1} \xi_{g_2}]$$

Also if we consider

$$g_N = 2 \sum_{|x| \le N} x\, \eta(x)$$

then

$$\mathcal{A}_N^0 g_N = \sum_{(x, x+1) \in [-N, N]} [\eta(x) - \eta(x+1)] = [\eta(-N) - \eta(N)]$$

and for $f = \eta(0) - \eta(1)$

$$\lim_{\substack{N \to \infty \\ \frac{n}{2N+1} \to \rho}} \frac{1}{2N} \sigma_N^2(n, f) = \lim_{\substack{N \to \infty \\ \frac{n}{2N+1} \to \rho}} \frac{1}{2N} D_{N,n}(g_N) = E^{\mu_\rho}[\xi_d^2]$$

with $\xi_d = 2(\eta(0) - \eta(1))$. In fact by poalrization we also have

$$\lim_{\substack{N \to \infty \\ \frac{n}{2N+1} \to \rho}} \frac{1}{2N} Cov_N(n, \mathcal{A}^0 g, f) = E^{\mu_\rho}[\xi_g \xi_d]$$

We can write any $f \in \mathcal{F}$ as a finite sum

$$\sum [u_a(\eta^{a,a+1}) - u_a(\eta)]$$

and let us assume that we have $f_0 = u_0(\eta^{a,a+1}) - u_0(\eta)$ for some a and $u_0 \in \mathcal{G}$. We shall compute

$$\lim_{\substack{N \to \infty \\ \frac{n}{2N+1} \to \rho}} \frac{1}{2N} Cov_N(n, f_0, f) = E^{\mu_\rho}[\xi_0 \xi]$$

Where $\xi = \xi_g$ if $f = \mathcal{A}^0 g$ and $\xi = \xi_d = 2(\eta(0) - \eta(1)$ if $f = \eta(0) - \eta(1)$. This is not a hard calculation. Denoting by $G_N = \sum_{|x| \le N-k} \tau_x g$

$$\lim_{\substack{N \to \infty \\ \frac{n}{2N+1} \to \rho}} \frac{1}{2N} Cov_N(n, f_0, \mathcal{A}^0 g)$$

$$= \lim_{\substack{N \to \infty \\ \frac{n}{2N+1} \to \rho}} \frac{2}{2N} < -(\mathcal{A}_N^0)^{-1} \mathcal{A}_N^0 G_N \,, \sum_x [(\tau_x u_0)(\eta^{a+u, a+u+1}) - (\tau_x u_0)(\eta)] >_{N,n}$$

$$= \lim_{\substack{N \to \infty \\ \frac{n}{2N+1} \to \rho}} \frac{2}{2N} < - \sum_x [G_N(\eta^{a+x, a+x+1}) - G_N(\eta)] \,, u_0(\eta)] >_{N,n}$$

$$= E^{\mu_\rho}[(-2\tau_{-a} u_0) \xi_g]$$

In exactly the same way the choice of $G_N = \sum_{|x|\le N} x\,\eta(x)$ produces

$$\lim_{\substack{N\to\infty \\ \frac{n}{2N+1}\to\rho}} \frac{1}{2N} Cov_N(n, f_0, \eta(0)-\eta(1)) = E^{\mu_\rho}[(-2\tau_{-a}u_0)\xi_d]$$

If we replace $-2\tau_{-a}u_0$ with its projection $\xi_0 = \mathbf{P}[-2\tau_{-a}u_0]$ onto $\mathcal{H}$, we get

$$\lim_{\substack{N\to\infty \\ \frac{n}{2N+1}\to\rho}} \frac{1}{2N} Cov_N(n, f_0, f) = E^{\mu_\rho}[\xi_o\xi]$$

where $f \to \xi$ is a linear map from $\mathcal{F}$ into $\mathcal{H}$. Once we prove Theorem 6.2, the range of this linear map will be dense in $\mathcal{H}$ and we will have

$$\liminf_{\substack{N\to\infty \\ \frac{n}{2N+1}\to\rho}} \frac{1}{2N}\sigma_N^2(n, f_0) \ge E^{\mu_\rho}[\xi_0^2]$$

We now try to control

$$\limsup_{\substack{N\to\infty \\ \frac{n}{2N+1}\to\rho}} \frac{1}{2N}\sigma_N^2(n, f)$$

If v_N solves $\mathcal{A}_N^0 v_N = \sum_{|x|\le N-k}\tau_x f$

$$\begin{aligned}
D_{N,n}(v_N) &= -< v_N(\eta)\,, \sum_x [(\tau_x u_0)(\eta^{x+a,x+a+1}) - (\tau_x u_0)(\eta)] >_{N,n} \\
&= -\sum_x < [v_N(\eta^{x+a,x+a+1}) - v_N(\eta)]\,, (\tau_x u_0)(\eta) >_{N,n} \\
&\le \sqrt{2D_{N,n}(v_N)}\sqrt{\sum_x \|\tau_x u_0\|_{N,n}^2}
\end{aligned}$$

providing the estimate

$$D_{N,n}(v_N) \le 2(2N+1)\|u_0\|_{N,n}^2$$

and therefore

$$\limsup_{\substack{N\to\infty \\ \frac{n}{2N+1}\to\rho}} \frac{1}{2N}\sigma_N^2(n, f) \le 4E^{\mu_\rho}[u_0^2]$$

To improve this bound we use a duality argument. We shall only sketch the proof.

$$\begin{aligned}
&\frac{1}{\sqrt{2N}}\sigma_N(n, f) \\
&= [\frac{1}{2N}\sigma_N^2(n, f)]^{\frac{1}{2}} \\
&= \frac{2}{2N}\sup_{w: D_{N,n}(w)\le 2N} E^{\nu_{N,n}}\Big[\sum_x [(\tau_x u_0)(\eta^{x+a,x+a+1}) - (\tau_x u_0)(\eta)]w(\eta)\Big] \\
&= \frac{2}{2N}\sup_{w: D_{N,n}(w)\le 2N} E^{\nu_{N,n}}\Big[\sum_x (\tau_x u_0)(\eta)[w(\eta^{x+a,x+a+1}) - w(\eta)]\Big] \\
&= \frac{2}{2N} E^{\nu_{N,n}}\Big[\sum_x (\tau_x u_0)(\eta)[w_N(\eta^{x+a,x+a+1}) - w_N(\eta)]\Big]
\end{aligned}$$

for some W_N with $D_{N,n}(w_N) \leq 2N$. On the average $w_N(\eta^{y,y+1}) - w_N(\eta)$ has L_2 norm 1, and a weak limit of averages of suitably chosen long blocks of

$$\tau_{-y}[w_N(\eta^{y,y+1}) - w_N(\eta)]$$

will produce a $\xi \in \mathcal{H}$ such that $E^{\mu_\rho}[\xi^2] \leq 1$ and

$$\limsup_{\substack{N\to\infty \\ \frac{n}{2N+1}\to\rho}} \frac{1}{2N}\sigma_N^2(n,f) = |E^{\mu_\rho}[2(\tau_{-a}u_0)\xi]|^2 \leq E^{\mu_\rho}[|\mathbf{P}(2(\tau_{-a}u_0))|^2]$$

so that f can be represented by the projection of $-2\tau_{-a}u_0$ □

Theorem 6.2 *The span $\mathcal{H}_0$ of $\xi_f^{0,1}$ in $L_2(\mu)$ as f varies over all local functions is of codimension one in $\mathcal{H}$ and a complementary (orthogonal) one dimensional subspace is provided by the span of $\eta(0) - \eta(1)$.*

Proof Let us start with a ξ in $\mathcal{H}$. We denote the translations $\tau_x\xi$ by $\xi^{x,x+1}$ and truncate them by taking conditional expectation given the σ-field generated by $\{\eta(x)\}$ in a finite interval $[-N,N]$ and get a $\xi_N^{x,x+1}$ that approximates $\xi^{x,x+1}$ quite well if $N - |x| >> 0$. We then solve simultaneously the equations

$$[v_N(\eta^{x,x+1}) - v_N(\eta)] = \xi_N^{x,x+1}(\eta) \quad \text{for all} \quad x$$

for all $(x, x+1) \in [-N,N]$, which can be done. If we fix a value n for the total number of particles in $[-N,N]$, then the equations can be 'integrated' upto a constant on each of the spaces $\Omega_{N,n}$ to a get solution v_N normalized to have mean 0 with repect to every uniform distribution $\nu_{N,n}$ on $\Omega_{N,n}$. We can get some estimates by thinking of v_N as the solution of the scalar (Poisson) equation.

$$\sum_{(x,x+1)\in[-N,N]} [v_N(\eta^{x,x+1}) - v_N(\eta)] = \sum_{(x,x+1)\in[-N,N]} \xi_N^{x,x+1}(\eta) \tag{6.1}$$

We get

$$\mathcal{D}_N(v_N) \leq CN$$

If we take $f_N = N^{-1}v_N$ and try with

$$F_N(\eta) = \sum_x \tau_x f_N$$

Then $F_N(\eta^{0,1}) - F_N(\eta)$ almost works except for some boundary terms. The two boundary terms are

$$f_N(\eta(0), \eta(2), \ldots, \eta(2N+1)) - f_N(\eta(1), \eta(2), \ldots, \eta(2N+1))$$

and a simlar one coming from the other end. Denote them by g_N. If we had an estimate of the form

$$\|g_N\|_2 \leq C$$

then the boundary terms will be bounded in $L_2(\mu)$ and yield a weak limit. From the estimates $D_{x,x+1}(v_N) \leq 1$ it is clear that $D_{x,x+1}(f_N) \to 0$ and this inturn can be shown to imply that any weak limit of g_N has to be of the form $c(\eta(0) - \eta(1))$ establishing that $\xi = c(\eta(0) - \eta(1))$ modulo $\mathcal{H}_0$.

Such an estimate does not exist, because the spectral gap on $[-N, N]$ for the operator appearing in (6.1) which is the generator of the symmetric simple exclusion process is cN^{-2} and the best we can do is

$$\|v_N\|_2^2 \le CN^3$$

There is no possibility of cancellation because the difference involves comparing the values of f_N on $\Omega_{N,n}$ for diffrent n and the Dirichlet form cannot help. We overcome the difficulty by solving (6.1) on $[-2N, 2N]$ to get v_{2N} and then integrating all the $\eta(x)$ outside $[-N, N]$ to obtain its conditional expectaion w_N defined on $\cup_n \Omega_{N,n}$. For w_N obtained in this manner Lemma 6.3 provides the estimate

$$\|w_N(\eta(0), \eta(2), \dots, \eta(2N+1)) - w_N(\eta(1), \eta(2), \dots, \eta(2N+1))\|_2 \le C$$

and this would complete the proof.

The orthogonality of ξ_f to $\eta(0) - \eta(1)$ is easily seen.

$$\begin{aligned}
&E^\mu\big[(F(\eta^{0,1}) - F(\eta))(\eta(0) - \eta(1))\big] \\
&= E^\mu\big[F(\eta^{0,1})\eta(0) - F(\eta^{0,1})\eta(1) - F(\eta)\eta(0) + F(\eta)\eta(1)\big] \\
&= E^\mu\big[F(\eta)\eta(1) - F(\eta)\eta(0) - F(\eta)\eta(0) + F(\eta)\eta(1)\big] \\
&= 2E^\mu\big[F(\eta)(\eta(1) - \eta(0))\big] \\
&= 2E^\mu\big[(F(\eta) - F(\tau\eta))\eta(1)\big] \\
&= 0
\end{aligned}$$

by the formal translation invariance of F. A rigorous proof involves truncating the sum that is F and passing to the limit. □

Lemma 6.3 *Let v_{2N} be the solution of*

$$\sum_{(x,x+1)\in[-2N,2N]} [v_{2N}(\eta^{x,x+1}) - v_{2N}(\eta)] = \sum_{(x,x+1)\in[-2N,2N]} \xi_{2N}^{x,x+1}$$

Define w_N by

$$w_N(\eta) = E^\mu\Big[w_{2N}(\eta)|\{\eta(x)\} : |x| \le N\Big]$$

Then

$$\|w_N(\eta(0), \eta(2), \dots, \eta(2N+1)) - w_N(\eta(1), \eta(2), \dots, \eta(2N+1))\|_2 \le C$$

Proof The main step consists of the following idea. We have two functions $f(0, \xi_1, \dots, \xi_n)$ and $f(1, \xi_1, \dots, \xi_n)$ and need to estimate

$$\sum_\xi [f(1, \xi_1, \dots, \xi_n) - f(0, \xi_1, \dots, \xi_n)]\rho^k(1-\rho)^{n-k}$$

where $k = k(\xi)$ is the number of 1's among $\xi_1, \dots, \xi_n$. We rewrite this as

$$\sum_k \binom{n}{k} \rho^k (1-\rho)^{n-k}[a_k - b_k]$$

where

$$a_k = \frac{1}{\binom{n}{k}} \sum_{k(\xi)=k} f(1, \xi_1, \dots, \xi_n)$$

and

$$b_k = \frac{1}{\binom{n}{k}} \sum_{k(\xi)=k} f(0, \xi_1, \ldots, \xi_n)$$

While one cannot estimate $|a_k - b_k|$ it is relatively easy to estimate $|a_k - b_{k+1}|$ because this involves interchanging a 0 and 1 and is controlled by the Dirichlet form. In fact we can write

$$|a_k - b_{k+1}| \leq \frac{1}{n\binom{n-1}{k-1}} \sum_{\substack{k(\xi)=k-1 \\ 1\leq i\leq n}} |f(0, \xi_1, \ldots, \xi_{i-1}, 1, \xi_{i+1}, \ldots, \xi_n)$$

$$- f(1, \xi_1, \ldots, \xi_{i-1}, 0, \xi_{i+1}, \ldots, \xi_n)|$$

which leads quickly to the kind of estimate that we need.

We now return to complete the proof of Theorem 6.1. The second statement is really the key. The final statement is a routine improvement of the second that we will not prove.

There are three different generators that we will consider. The generator $\mathcal{A}$ that is our primary concern, its symmetrized version $\mathbf{A}$ and the generator of the symmetric nearest neighbor exclusion process $\mathcal{A}^0$

For any local function $f \in \mathcal{F}$ we associated a vector $\xi = Tf \in \mathcal{H}$ such that

$$\lim_{\substack{N\to\infty \\ \frac{n}{2N}\to\rho}} \frac{1}{2N}\sigma^2_{\mathcal{A}^0}(N,n,f) = E^\rho[\|\xi\|^2]$$

The same method would have yielded a similar result for $\mathbf{A}$, with $\xi = T_{\mathbf{A}}f$

$$\lim_{\substack{N\to\infty \\ \frac{n}{2N}\to\rho}} \frac{1}{2N}\sigma^2_{\mathbf{A}}(N,n,f) = E^\rho[\|\xi\|^2_{\mathbf{A}}]$$

where

$$\|\xi\|^2_{\mathbf{A}} = \sum p(z)\|\xi_{0,z}\|^2$$

and some elementary algebra allows one to express $\xi_{0,z}$ in terms of $\xi = \xi_{0,1}$. Formally $f = \mathbf{A}g$, $F = \sum_x \tau_x g$ and

$$\xi_{0,x} = F(\eta^{0,x}) - F(\eta)$$

Because of the equivalence of the Dirichlet Forms corresponding to $\mathcal{A}^0$ and $\mathbf{A}$, $E^\rho[\|\xi\|^2]$ and $E^\rho[\|\xi\|^2_{\mathbf{A}}]$ are equivalent. We need to prove that for some c

$$\inf_g E^\rho[|T_{\mathbf{A}}(W - \mathcal{A}g - c(\eta(1) - \eta(0)))|^2] = 0$$

which would follow if we show that the closure in $\mathcal{H}$ of $\{T_{\mathbf{A}}\mathcal{A}g\}$ as g varies over local functions is of codimension 1 and does not contain $\eta(1) - \eta(0)$. The similar statement with $\{T_{\mathbf{A}}\mathbf{A}g\}$ replacing $\{T_{\mathbf{A}}\mathcal{A}g\}$, which we proved for $\mathcal{A}^0$ can be established in the same manner. The orthogonality of $T_{\mathbf{A}}\mathbf{A}g$ and $\eta(1) - \eta(0)$ will now hold for the inner product in the Hilbert space with norm $E^\rho[\|\xi\|^2_{\mathbf{A}}]$.

Let us consider the family $\mathcal{A}_t = t\mathcal{A} + (1-t)\mathbf{A}$ that corresponds to simpe exclusion with $p_t(z) = \frac{1+t}{2}p(z) + \frac{1-t}{2}p(-z)$. Denoting by $\mathcal{H}_t$ the range of $\{T_{\mathbf{A}}(\mathcal{A}_t g + c(\eta(1) - \eta(0))\}$ as g and c vary, we can map $\mathcal{H} \to \mathcal{H}_t$ by the linear map S_t sending $T_{\mathbf{A}}(\mathcal{A}_0 g + c(\eta(1) - \eta(0)) \to T_{\mathbf{A}}(\mathcal{A}_t g + c(\eta(1) - \eta(0))$. Note that $\mathcal{A}_0 = \mathbf{A}$. The boundedness of $\mathcal{A}$ from $H_1 \to H_{-1}$ makes this a bounded family of perturbations in the uniform norm. Since $\mathcal{H}_0$ is dense in $\mathcal{H}$, S_t extends as continuous perturbation of $S_0 = I$. If we show that $\|S_t\xi\| \geq C\|\xi\|$ uniformly in $0 \leq t \leq 1$ for some $C > 0$, we will succeed in our goal.

Let

$$u_t = c(\eta(1) - \eta(0)) + \mathcal{A}_t g$$

We need a uniform lower bound on $<< u_t, u_t >>_\rho$. From the variational formula we see that

$$\begin{aligned} << u_t, u_t >>_\rho &\geq \lim_{N\to\infty} \frac{1}{2N} E\Big[\sum_{|x|\leq N-k} \tau_x u_t \cdot \sum_{|x|\leq N-k} \tau_x g - \frac{1}{2} D\big(\sum_{|x|\leq N-k} \tau_x g\big)\Big] \\ &= \lim_{N\to\infty} \frac{1}{2N} E\Big[\mathcal{A}_t \sum_{|x|\leq N-k} \tau_x g \cdot \sum_{|x|\leq N-k} \tau_x g - \frac{1}{2} D\big(\sum_{|x|\leq N-k} \tau_x g\big)\Big] \\ &= \frac{1}{2} \lim_{N\to\infty} \frac{1}{2N} D\big(\sum_{|x|\leq N-k} \tau_x g\big) \\ &\geq \delta << \mathbf{A}g, \mathbf{A}g >>_\rho \\ &= \delta << u_0 - c(\eta(1) - \eta(0)), u_0 - c(\eta(1) - \eta(0)) >>_\rho \end{aligned}$$

for some $\delta > 0$. From $\|S_t x + y\| \geq \delta\|x\|$ one easily obtains $\|S_t x + y\| \geq C_\delta\|x + y\|$. Take $x = u_0$ and $y = c(\eta(1) - \eta(0))$.

References

[1] Kipnis, C., and Landim, C. [1999] *Scaing Limits of Interacting Particle Systems*, Springer-Verlag, Heidelberg.

[2] Kipnis, C., Olla, S., and Varadhan, S. R. S. [1989] *Hydrodynamics and large deviation for simple exclusion process*, Comm. Pure Appl. Math. **42**, 115–137.

[3] Lin, Xu. [1993] *Hydrodynamics for asymmetric mean zero simple exclusion, Ph.D. Thesis, New York University.*

[4] Quastel, J. [1992] *Diffusion of color in simple exclusion process*, Comm. Pure Appl. Math. **45**, 623–680.

[5] Yau, H. T. [1991] *Relative entropy and the hydrodynamics of Ginzburg–Landau models*, Lett. Math. Phys. **22**, 63–80.

Part 2

Fields Institute Communications
Volume **27**, 2000

On a 1-D Granular Media Immersed in a Fluid

J. A. Carrillo
Department of Mathematics
University of Texas at Austin
Austin-Texas, 78712 USA
carri@math.utexas.edu

Abstract. In this paper a 1-D model for the evolution of inelastic particles in a thermal bath is studied. We derive an equation for the evolution of the probability density of the particles in phase space. The analysis of the homogeneous equation is the main objective of this paper. The existence and uniqueness of this equation is addressed. Finally, we prove that the asymptotic behavior is governed by a stationary solution of the homogeneous equation. A formal hydrodynamic set of equations is derived for the non-homogeneous case.

1 Introduction

Let us consider a system of N heavy particles in a one dimensional setting. We assume that these particles are surrounded by a thermal reservoir or a fluid of small particles which produces on them a stochastic force (Brownian motion). Also, we take into account the collisions between the heavy particles that we assume of inelastic type.

Let x_i, $v_i \in \mathbb{R}$ be the position and the velocity of the i-th particle. The dynamics of this system is as follows. The particles move under the action of the stochastic force between collisions. Once two particles are in the same position, they collide according to the rule:

$$v' = v_1 + \epsilon(v - v_1), \qquad v_1' = v - \epsilon(v - v_1)$$

where v', v_1', v and v_1 are the outgoing and ingoing velocities respectively and ϵ is a real parameter measuring the degree of inelasticity of the collision. Since the particles are all assumed to be identical, we can substitute the previous rule of collision by

$$v' = v - \epsilon(v - v_1), \qquad v_1' = v_1 + \epsilon(v - v_1). \tag{1.1}$$

This change is made only to simplify some computations.

2000 *Mathematics Subject Classification.* Primary 35A05, 35B40; Secondary 82C40, 35Q99.
Partially supported by the DGES MEC-Spain Perfeccionamiento de Doctores en el Extranjero fellowship, TMR European Union ERB FMBX-CT97-0157 and DGES (Spain) PB98-1281.

As a consequence, the Langevin stochastic-differential system governing the time evolution of the system is:

$$x_i' = v_i, \quad v_i' = -\beta v_i + \Gamma(t) + \epsilon \sum_{j=1}^{N} \delta(x_i - x_j)(v_j - v_i)|v_j - v_i| \tag{1.2}$$

where $\beta > 0$ is the friction parameter, $\Gamma(t)$ is a Gaussian stochastic process with zero mean and variance $\sigma > 0$, $\epsilon(v_j - v_i)$ is the jump performed by the particle i after a collision with the particle j due to (1.1) and $\delta(x_i - x_j)|v_j - v_i| = \delta(t - t_{i,j})$, being $t_{i,j}$ the instant of impact between the particle i and j.

Using Ito's stochastic integral theorem to (1.2), the equation for the time evolution of the density $\mu^N(t, x_1, \ldots, x_N, v_1, \ldots, v_N)$ is given by

$$\partial_t \mu^N + \sum_{i=1}^{N} v_i \partial_{x_i} \mu^N = \sum_{i=1}^{N} \left(\beta \partial_{v_i}(v_i \mu^N) + \sigma \partial^2_{v_i} \mu^N\right) - \epsilon \sum_{i \neq j} \delta(x_i - x_j) \partial_{v_i}(\phi(v_j - v_i)\mu^N)$$

where $\phi(\bar{v} - v) = (\bar{v} - v)|\bar{v} - v|$. Proceeding as in the BBGKY-hierarchy, assuming the hypothesis of propagation of chaos and performing the scaling limit $\epsilon \to 0$, $N \to \infty$ in such a way that $N\epsilon \to \lambda > 0$ (see [2]), we conclude that the time evolution for the one particle distribution function is given by the kinetic equation

$$\partial_t f + v \partial_x f + \lambda \partial_v (Ff) - \beta \partial_v (vf) - \sigma \partial_v^2 f = 0 \quad \text{in } (0, T) \times \mathbb{R}_x \times \mathbb{R}_v \tag{1.3}$$

where

$$F(t, x, v) = \int_{\mathbb{R}} \phi(\bar{v} - v) f(t, x, \bar{v}) \, d\bar{v}. \tag{1.4}$$

The case in which there is no surrounding thermal bath, i. e., $\beta = \sigma = 0$, has been considered previously in [2]. The stochastic differential system (1.2) has been analyzed in [1]. In fact, inelastic collapse, infinite number of collisions in a finite time, exists for a three particle system. The main questions are whether the thermal bath drives the particles to a stationary state and if singularities in the non-homogeneous equation can exist. We prove the existence of smooth solutions for the homogeneous equation driven by the thermal bath to a steady state.

In this paper we aim to the analysis of the homogeneous solutions of the nonlinear integro-differential equation (1.3), that is,

$$\partial_t f + \lambda \partial_v (Ff) - \beta \partial_v (vf) - \sigma \partial_v^2 f = 0 \quad \text{in } (0, T) \times \mathbb{R} \tag{1.5}$$

with $F(t, v) = (-\phi) * f(t)$. In the sequel, we refer to this equation as the granular Fokker-Planck (GFP) equation.

The asymptotic behavior of equation (1.5) has been studied in [3]. Here, we focus on the existence and uniqueness of (1.5) (Sections 2 and 3) and in deriving a formal hydrodynamics around the stationary solution of (1.5) in the non-homogeneous equation (1.3) (Section 4). The main result of this paper is the following

Theorem 1.1 *The equation* (1.5) *with initial data* $(1 + v^4) f_o \in L^1(\mathbb{R})$ *has a unique solution with the regularity* $(1 + v^2)^{-1} F \in L^\infty((0, T) \times \mathbb{R})$ *and* $(1 + v^4) f \in C([0, T], L^1(\mathbb{R}))$ *for any* $T > 0$. *Moreover the solution verifies:*

1. $F \in C^1([0, T] \times \mathbb{R})$ *for any* $T > 0$.
2. $(1 + v^4) f$ *is bounded in* $L^1(\mathbb{R})$ *independently of* $t \geq 0$.
3. *The solution verifies the integral equation*

$$f(t, v) = (M_{\nu(t)} * f_o)_{\alpha(t)} + \lambda \int_0^t e^{\beta s} \left(-\partial_v M_{\nu(t-s)} * (F(s) f(s))_{\alpha(s)^{-1}}\right)_{\alpha(t)} ds.$$

4. *If in addition* $f_o \in C^2(\mathbb{R})$, *then the solution is classical.*

2 Linear Fokker-Planck equation

Let us consider the linear Fokker-Planck equation equation

$$\partial_t f + \lambda \partial_v (F_o f) - \beta \partial_v (v f) - \sigma \partial_v^2 f \; = \; 0 \quad \text{in } (0,T) \times \mathbb{R} \tag{2.1}$$

where $F_o(t,v) \in L^\infty(0,T; W^{2,\infty}(\mathbb{R}))$, for any $T > 0$.

The fundamental solution $G(t,s,v,w)$, $t,s \in \mathbb{R}^+$, $v,w \in \mathbb{R}$ of this equation with $F_o = 0$ is the well-known kernel of the adjoint Ornstein-Uhlenbeck stochastic process. It can be easily computed using Fourier transform in v and solving the obtained hyperbolic equation through characteristics and is given by

$$G(t,s,v,w) \; = \; \alpha(t)^{-1/2} \, M_{\nu(t-s)} \left(\alpha(t)^{-1/2} v - \alpha(s)^{-1/2} w \right)$$

where $\alpha(t) \; = \; e^{-2\beta t}$, $\nu(t) \; = \; \frac{\sigma}{\beta}(e^{2\beta t} - 1)$ and M_α with $\alpha > 0$ is given by

$$M_\alpha(v) \; = \; (2\pi\alpha)^{-1/2} \exp\left\{ -\frac{|v|^2}{2\alpha} \right\}.$$

Thus, the solution of (2.1) with $F_o = 0$ and initial data $f_o \in L^1(\mathbb{R})$ can be written in convolution form as

$$f(t,v) \; = \; (M_{\nu(t)} * f_o)_{\alpha(t)} \; = \; M_{\gamma(t)} * (f_o)_{\alpha(t)},$$

where $\gamma(t) \; = \; \frac{\sigma}{\beta}(1 - e^{-2\beta t})$ and

$$(g)_\alpha(v) \; = \; \alpha^{-1/2} \, g(\alpha^{-1/2} v) \;, \; v \in \mathbb{R}, \; \alpha > 0$$

and any function g. In the same way, we can solve the initial value problem for equation (2.1) with initial data $f_o \in L^1(\mathbb{R})$ and right-hand side $h \in C([0,T], L^1(\mathbb{R}))$ for any $T > 0$. The solution is given by

$$f(t,v) \; = \; (M_{\nu(t)} * f_o)_{\alpha(t)} \; + \; \int_0^t (M_{\nu(t-s)} * (h(s))_{\alpha(s)^{-1}})_{\alpha(t)} \, ds.$$

G is positive and it satisfies

$$\int_{\mathbb{R}} G(t,s,v,w) \, dw \; = \; e^{\beta(t-s)}$$

and

$$\int_{\mathbb{R}} G(t,s,v,w) \, dv \; = \; 1,$$

for any $t,s > 0$, $v,w \in \mathbb{R}$. It is well known that

$$\|\partial_v^k (M_\alpha * g)\|_{L^q(\mathbb{R})} \; \le \; C \, \gamma(t)^{-\frac{1}{2}(\frac{1}{p} - \frac{1}{q}) - \frac{k}{2}} \, \|g\|_{L^p(\mathbb{R})} \tag{2.2}$$

for any $1 \le p \le q \le \infty$ and any natural k. Also, it is straightforward to check that

$$\|(g)_\alpha\|_{L^p(\mathbb{R})} \; = \; \alpha(t)^{-\frac{1}{2p'}} \, \|g\|_{L^p(\mathbb{R})}$$

for any $1 \le p \le \infty$. As a consequence, we have the following result, see [8] for more properties of this equation.

Lemma 2.1 *Given $f_o \in L^1(\mathbb{R})$ positive, $1 \le p \le q \le \infty$ and any natural k, there exists a constant C such that the unique solution $f(t,v)$ of (2.1) with $F_o = 0$ satisfies*

$$\|\partial_v^k f(t,\cdot)\|_{L^q(\mathbb{R})} \;\le\; C\;\gamma(t)^{-\frac{1}{2}(\frac{1}{p}-\frac{1}{q})-\frac{k}{2}}\;\alpha(t)^{-\frac{1}{2p'}}\;\|f_o\|_{L^p(\mathbb{R}^N)}$$

In the general case we look for the fundamental solution G_{F_o} of the equation (2.1). Formally, $G_{F_o}(t,s,v,w)$ verifies

$$G_{F_o}(t,s,v,w) \;=\; G(t,s,v,w) +$$

$$\lambda \int_s^t \int_{\mathbb{R}} \partial_{w'} G(t,s',v,w')\; F_o(s',w')\; G_{F_o}(s',s,w',w)\; dw'\; ds'.$$

The existence of $G_{F_o}(t,s,v,w)$ is proved using an iterative scheme in the previous formula as in [12]. More exactly, we write the following theorem without proof since it can be deduced with a procedure totally analogous to [12].

Theorem 2.2 *There exists a fundamental solution $G_{F_o}(t,s,v,w)$ of equation* (2.1) *satisfying:*

1. *G_{F_o} is positive and*
$$\int_{\mathbb{R}} G_{F_o}(t,s,v,w)\; dv \;=\; 1,$$
for any $t > s > 0$, $v,w \in \mathbb{R}$.
2. *There exists a constant $C = C(F_o) > 0$ depending exponentially on the norm of F_o in $L^\infty(0,T;W^{2,\infty}(\mathbb{R}))$ such that*
$$G_{F_o}(t,s,v,w) \;\le\; C\; G(t,s,\frac{v}{2},\frac{w}{2})$$
and
$$|\partial_v G_{F_o}(t,s,v,w)| \;\le\; C\;(t-s)^{-\frac{1}{2}}\; G(t,s,\frac{v}{2},\frac{w}{2})$$
for any $t > s > 0$, $v,w \in \mathbb{R}$.
3. *Given $f_o \in L^1(\mathbb{R}) \cap C_b(\mathbb{R})$, there exists a unique classical solution of* (2.1) *given by*
$$f(t,v) \;=\; \int_{\mathbb{R}} G_{F_o}(t,0,v,w)\; f_o(w)\; dw.$$
Also, the solution verifies
$$f(t,v) \;=\; (M_{\nu(t)} * f_o)_{\alpha(t)} \;+\; \lambda \int_0^t e^{\beta s}\;(-\partial_v M_{\nu(t-s)} * (F_o(s)f(s))_{\alpha(s)^{-1}})_{\alpha(t)}\; ds.$$
4. *Given F_1, F_2 in $L^\infty(0,T;W^{2,\infty}(\mathbb{R}))$ for any $T > 0$, there exists a constant C depending exponentially on the norms of F_1 and F_2, such that*
$$|G_{F_1}(t,s,v,w) - G_{F_2}(t,s,v,w)| \;\le\; C\;\|F_1 - F_2\|_{L^\infty(0,T;W^{2,\infty}(\mathbb{R}))}\; G(t,s,\frac{v}{2},\frac{w}{2})$$
for any $T > t > s > 0$, $v,w \in \mathbb{R}$.

3 Nonlinear regularized GFP equation

In order to solve the nonlinear equation (1.5) we can try to apply classical iterative procedures and a fixed point theorem. Here, $\bar{f}$ stands for $f(t,\bar{v})$. However, if we compute formally the equation for the evolution of the kinetic energy, we have

$$\frac{d}{dt}\int_{\mathbb{R}} v^2 f\, dv = -2\beta\int_{\mathbb{R}} v^2 f\, dv + 2\sigma M + 2\lambda\int_{\mathbb{R}^2}\phi(\bar{v}-v)vf\bar{f}\, dv\, d\bar{v},$$

where M is the mass of the initial data. Using the antisymmetry of ϕ we have that

$$2\int_{\mathbb{R}^2}\phi(\bar{v}-v)vf\bar{f}\, dv\, d\bar{v} = -\int_{\mathbb{R}^2}|\bar{v}-v|^3 f\bar{f}\, dv\, d\bar{v} \leq 0$$

and thus, formally the kinetic energy for the model is uniformly bounded for any $t\geq 0$. This argument can not be carried out if the force field F is defined by a different f to perform a fixed point argument. Therefore, in order to keep the control over the kinetic energy we are going to prove the existence of the solution for a regularized GFP equation using a fixed point argument. Since we will be allowed at this stage to perform the previous computations due to the smoothness of the solution, we will able then, to pass to the limit. This problem appears in related equations [6, 7, 10, 11, 12].

The aim of this section is to prove the existence for the GFP$^\epsilon$ problem (3.1). Let us consider ϕ^ϵ defined by

$$\phi^\epsilon(v) = vAbs^\epsilon(v)\mathcal{X}^\epsilon(v)$$

with $\mathcal{X}^\epsilon\in C_o^\infty(\mathbb{R})$ such that

$$\begin{aligned}
\mathcal{X}^\epsilon(v) &= 0 \text{ for } |v|\geq\frac{2}{\epsilon},\\
\mathcal{X}^\epsilon(v) &= 1 \text{ for } |v|\leq\frac{1}{\epsilon},\\
\operatorname{sign}(v)\,\partial_v\mathcal{X}^\epsilon(v) &< 0 \text{ for } \frac{1}{\epsilon}<|v|<\frac{2}{\epsilon}\\
\mathcal{X}^\epsilon(v) &= \mathcal{X}^\epsilon(-v) \text{ and } |\partial_v\mathcal{X}^\epsilon(v)| \leq C\,\epsilon\,,\ \forall v\in\mathbb{R},
\end{aligned}$$

and $Abs^\epsilon(v)$ is a smooth approximation of $|v|$ converging uniformly to $|v|$ as $\epsilon\to 0$, symmetric $Abs^\epsilon(v) = Abs^\epsilon(-v)$, $\forall v\in\mathbb{R}$ and $0\leq Abs^\epsilon(v)\leq|v|$, $\forall v\in\mathbb{R}$.

The GFP$^\epsilon$ equation is given by

$$\partial_t f^\epsilon + \lambda\partial_v(F^\epsilon f^\epsilon) - \beta\partial_v(vf^\epsilon) - \sigma\partial_v^2 f^\epsilon = 0 \quad \text{in } (0,T)\times\mathbb{R} \tag{3.1}$$

with $F^\epsilon(t,v) = (-\phi^\epsilon)*f^\epsilon(t)$ and initial condition f_o^ϵ where f_o^ϵ is a regularization of compact support of an initial data $(1+v^4)f_o\in L^1(\mathbb{R})$ with the same mass as f_o and bounded moments up to order 4 in $L^1(\mathbb{R})$ uniformly on ϵ.

We are going to prove the existence of a weak solution for equation (3.1) using a fixed point argument. Since equation (3.1) is mass-preserving, it is easy to see that

$$\|F^\epsilon\|_{L^\infty(0,T;W^{k,p}(\mathbb{R}))} \leq B(k,p,\epsilon)\|f_o\|_{L^1(\mathbb{R})}$$

for any $k\in\mathbb{N}$ and $1\leq p\leq\infty$, where B is a constant depending on k, p and ϵ.

Let us define by $\mathcal{F}^\epsilon$ the set of positive functions f in $C([0,T],L^1\cap C_b(\mathbb{R}))$ with $f(0,v)=f_o^\epsilon$ and

$$\|f(t)\|_{L^1(\mathbb{R})} = \|f_o^\epsilon\|_{L^1(\mathbb{R})}$$

and

$$\|f\|_{L^\infty((0,T)\times\mathbb{R})} \leq A(T,\epsilon)\,\|f_o^\epsilon\|_{L^\infty(\mathbb{R})}$$

for any $t \geq 0$ and any $T > 0$, where $A(T, \epsilon)$ depends only on f_o and the physical constants σ and β. In fact, since for any F_o as in theorem 2.2 we have

$$G_{F_o}(t, s, v, w) \leq C(F_o)\, G(t, s, \frac{v}{2}, \frac{w}{2}).$$

Thus, we can obtain the a priori bound

$$\left| \int_{\mathbb{R}} G_{F_o}(t, 0, v, w)\, f_o(w)\, dw \right| \leq C(F_o) \int_{\mathbb{R}} G(t, s, \frac{v}{2}, \frac{w}{2})\, f_o(w)\, dw$$

$$\leq C(F_o)\, e^{\beta t}\, \|f_o^\epsilon\|_{L^\infty(\mathbb{R})}.$$

Therefore, let us define $A(T, \epsilon)$ by

$$A(T, \epsilon) = C(B(2, \infty, \epsilon)\|f_o\|_{L^1(\mathbb{R})})\, e^{\beta T}.$$

In this way, we define the map φ^ϵ from $\mathcal{F}^\epsilon$ into $C([0, T], L^1 \cap C_b(\mathbb{R}))$ given by $\varphi^\epsilon(g) = f$ where $f(t, v)$ is the solution of (2.1) with $F_o = F_g^\epsilon = (-\phi^\epsilon) * g$. It is straightforward to see using theorem 2.2 that $\varphi^\epsilon(\mathcal{F}^\epsilon) \subset \mathcal{F}^\epsilon$ and that $\mathcal{F}^\epsilon$ is a closed, convex, bounded subset of the Banach space $X = C([0, T], L^1 \cap C_b(\mathbb{R}))$. Moreover, given $g_1, g_2 \in \mathcal{F}^\epsilon$ we have

$$(F_{g_1}^\epsilon - F_{g_2}^\epsilon)(t, v) = \int_{\mathbb{R}} \phi^\epsilon(\bar{v} - v)(g_1 - g_2)(t, \bar{v})\, d\bar{v},$$

and thus,

$$\|F_{g_1}^\epsilon - F_{g_2}^\epsilon\|_{L^\infty(0,T;W^{k,p}(\mathbb{R}))} \leq C(k, p, \epsilon)\, \|g_1 - g_2\|_X.$$

Therefore, using theorem 2.2 we have that

$$|G_{F_{g_1}^\epsilon}(t, s, v, w) - G_{F_{g_2}^\epsilon}(t, s, v, w)|$$

$$\leq C(B(2, 1, \epsilon)\|f_o\|_{L^1(\mathbb{R})})\, \|F_{g_1}^\epsilon - F_{g_2}^\epsilon\|_{L^\infty(0,T;W^{2,\infty}(\mathbb{R}))}\, G(t, s, \frac{v}{2}, \frac{w}{2})$$

and as a consequence

$$\|\varphi^\epsilon(g_1) - \varphi^\epsilon(g_2)\|_{C([0,T],L^1(\mathbb{R}))} \leq C(\epsilon)\, \|F_{g_1}^\epsilon - F_{g_2}^\epsilon\|_{L^\infty(0,T;W^{2,\infty}(\mathbb{R}))}\, \|f_o\|_{L^1(\mathbb{R})}$$

and

$$\|\varphi^\epsilon(g_1) - \varphi^\epsilon(g_2)\|_{C([0,T],C_b(\mathbb{R}))} \leq C(\epsilon)\, \|F_{g_1}^\epsilon - F_{g_2}^\epsilon\|_{L^\infty(0,T;W^{2,\infty}(\mathbb{R}))}\, \|f_o^\epsilon\|_{L^\infty(\mathbb{R})}\, e^{\beta T}.$$

We conclude that φ^ϵ is continuous.

Finally, in order to apply Schauder's fixed point theorem we need to prove the relative compactness of the image of φ^ϵ. Using again theorem 2.2 we have

$$|\partial_v G_{F_g^\epsilon}(t, s, v, w)| \leq C(B(2, 1, \epsilon)\|f_o\|_{L^1(\mathbb{R})})\, (t - s)^{-\frac{1}{2}}\, G(t, s, \frac{v}{2}, \frac{w}{2})$$

and thus, $\partial_v \varphi^\epsilon(g)$ is bounded uniformly on g for any $t > 0$. Therefore, the set $\{f(t), f \in \varphi^\epsilon(\mathcal{F}^\epsilon)\}$ is relatively compact in $C(K)$ for any K compact set on $\mathbb{R}$. In the same way, we have that

$$\|\partial_v \varphi^\epsilon(g)(t)\|_{L^1(\mathbb{R})} \leq C(B(2, 1, \epsilon)\|f_o\|_{L^1(\mathbb{R})})\, t^{-\frac{1}{2}}$$

for any $t > 0$. Therefore, using Frechet-Kolmogorov's theorem, the set $\{f(t), f \in \varphi^\epsilon(\mathcal{F}^\epsilon)\}$ is relatively compact in $L^1(K)$ for any K compact set on $\mathbb{R}$. To obtain the compactness in $L^1(\mathbb{R}) \cap C_b(\mathbb{R})$ it suffices to show that for any $r > 0$, there exists $R > 0$ such that

$$\int_{\mathbb{R} - B(0,R)} \varphi^\epsilon(g)(t, v)\, dv < r \tag{3.2}$$

for any $g \in \mathcal{F}^\epsilon$. This fact is classically known and is a consequence of the equicontinuity of the set and the versions of the Ascoli-Arzela and Frechet-Kolmogorov theorems in $\mathbb{R}$.

Let us now prove (3.2). It also suffices to show that the kinetic energy is uniformly bounded in the set $\{f(t), f \in \varphi^\epsilon(\mathcal{F}^\epsilon)\}$. Taking into account that the solutions $\varphi^\epsilon(g)$ are classical, we deduce

$$\frac{d}{dt}\int_{\mathbb{R}} v^2\varphi^\epsilon(g)\,dv \;=\; -2\beta\int_{\mathbb{R}} v^2\varphi^\epsilon(g)\,dv \;+\; 2\sigma M + 2\lambda\int_{\mathbb{R}} vF_g^\epsilon\varphi^\epsilon(g)\,dv.$$

We can bound the last term of the right-hand side as

$$\begin{aligned}\left|\int_{\mathbb{R}} vF_g^\epsilon\varphi^\epsilon(g)\,dv\right| \;&\leq\; B(2,1,\epsilon)\|f_o\|_{L^1(\mathbb{R})}\int_{\mathbb{R}}|v|\varphi^\epsilon(g)\,dv\\ &\leq\; C(\epsilon)\left(\|f_o\|_{L^1(\mathbb{R})}+\int_{\mathbb{R}} v^2\varphi^\epsilon(g)\,dv\right).\end{aligned}$$

Thus,

$$\frac{d}{dt}\int_{\mathbb{R}} v^2\varphi^\epsilon(g)\,dv \;\leq\; 2\sigma M + \;C(\epsilon)\left(M+\int_{\mathbb{R}} v^2\varphi^\epsilon(g)\,dv\right)$$

for any $t>0$, which implies that the kinetic energy of $\varphi^\epsilon(g)$ is bounded in finite time intervals uniformly on g.

Summarizing the last paragraphs we have proved that the set $\{f(t), f \in \varphi^\epsilon(\mathcal{F}^\epsilon)\}$ is relatively compact in $L^1 \cap C_b(\mathbb{R})$ for any time $t \geq 0$. To finish the proof of the compactness of the set $\varphi^\epsilon(\mathcal{F}^\epsilon)$ on X we must show the equicontinuity in time. Theorem 2.2 implies that for any $g \in \mathcal{F}^\epsilon$ we have

$$\varphi^\epsilon(g)(t,v)-\varphi^\epsilon(g)(s,v) \;=\; \lambda\int_s^t e^{\beta s'}\;(-\partial_v M_{\nu(t-s')}*(F_g^\epsilon(s')\varphi^\epsilon(g)(s'))_{\alpha(s')^{-1}})_{\alpha(t)}\,ds'$$

for any $0 \leq s \leq t$. Therefore, we obtain that

$$\begin{aligned}&\|\varphi^\epsilon(g)(t)-\varphi^\epsilon(g)(s)\|_{L^1(\mathbb{R})}\\ &\leq\;\lambda\int_s^t e^{\beta s'}\;\nu(t-s')^{-\frac{1}{2}}\;\|F_g^\epsilon(s')\|_{L^\infty(\mathbb{R})}\;\|f_o\|_{L^1(\mathbb{R})}\,ds'\end{aligned} \tag{3.3}$$

and

$$\begin{aligned}&\|\varphi^\epsilon(g)(t)-\varphi^\epsilon(g)(s)\|_{C_b(\mathbb{R})}\\ &<\;\lambda\int_s^t e^{\beta s'}\;\nu(t-s')^{-\frac{1}{2}}\;\|F_g^\epsilon(s')\|_{L^\infty(\mathbb{R})}\;\|\varphi^\epsilon(g)(s')\|_{L^\infty(\mathbb{R})}\,ds'.\end{aligned} \tag{3.4}$$

Since F_g^ϵ is L^∞-bounded and $\varphi^\epsilon(g)$ are uniformly bounded independently of g for any $T > 0$, we have that

$$\|\varphi^\epsilon(g)(t)-\varphi^\epsilon(g)(s)\|_X \;\leq\; C\,(\Phi(t)-\Phi(s))$$

where

$$\Phi(t) \;=\; \int_0^t e^{\beta s'}\;\nu(t-s')^{-\frac{1}{2}}\,ds'.$$

Since Φ is continuous, we have the uniform equicontinuity in time of the set $\varphi^\epsilon(\mathcal{F}^\epsilon)$ on $C([0,T], L^1 \cap C_b(\mathbb{R}))$ for any $T > 0$.

As a consequence, we can apply Schauder's fixed point theorem to deduce the existence of a solution f^ϵ of equation (3.1). The rest of the section is devoted to the analysis of the properties of this solution.

The first remark is that the solution is classical by using theorem 2.2. Moreover, the derivatives up to second order of f^ϵ are continuous with limit zero at infinity and integrable. Therefore, we can obtain easily by the divergence theorem the evolution for the moments of f^ϵ. Precisely, let us denote by $m_k(f^\epsilon)$ the moment in v of order k of the distribution f^ϵ, that is,

$$m_k(f^\epsilon) = \int_{\mathbb{R}} |v|^k f^\epsilon \, dv.$$

As a consequence, we have

$$\frac{d}{dt}\int_{\mathbb{R}} f^\epsilon \, dv = 0, \quad \frac{d}{dt}\int_{\mathbb{R}} v f^\epsilon \, dv = -\beta \int_{\mathbb{R}} v f^\epsilon \, dv$$

and

$$\frac{d}{dt} m_k(f^\epsilon) = -k\beta m_k(f^\epsilon) + k(k-1)\sigma m_{k-2}(f^\epsilon)$$
$$- \frac{k}{2}\lambda \int_{\mathbb{R}^2} \left(\bar{v}|\bar{v}|^{k-2} - v|v|^{k-2}\right) \phi^\epsilon(\bar{v} - v) \, f^\epsilon \bar{f}^\epsilon \, dv \, d\bar{v} \tag{3.5}$$

for any $k \geq 2$. Also, we obtain that the entropy $H(f^\epsilon)$ given by

$$H(f^\epsilon) = \int_{\mathbb{R}} f^\epsilon \log f^\epsilon \, dv$$

satisfies

$$\frac{d}{dt} H(f^\epsilon) = \beta - 4\sigma \int_{\mathbb{R}} \left|\partial_v \sqrt{f}\right|^2 dv + 2\lambda \int_{\mathbb{R}^2} \partial_v \phi^\epsilon(\bar{v} - v) \, f^\epsilon \bar{f}^\epsilon \, dv \, d\bar{v}. \tag{3.6}$$

Using the above production laws for the moments of the distribution we deduce the key estimates on the solution.

Lemma 3.1 *Given any $k \geq 0$ integer and any $t \geq 0$, we have*

$$m_{2k}(f^\epsilon) \leq A_{2k}$$

where

$$A_{2k} = m_{2k}(f_o^\epsilon) + (2k-1)\frac{\sigma}{\beta} A_{2k-2}, \quad k \geq 1$$

and $A_o = M$.

Proof .- The production laws (3.5) imply that

$$\frac{d}{dt} m_{2k}(f^\epsilon) \leq -2k\beta m_{2k}(f^\epsilon) + 2k(2k-1)\sigma m_{2k-2}(f^\epsilon)$$

for any $k \geq 1$. Therefore, we conclude by integration that

$$m_{2k}(f^\epsilon)(t) \leq e^{-2k\beta t} m_{2k}(f_o^\epsilon) + 2k(2k-1)\sigma e^{-2k\beta t} \int_0^t e^{2k\beta s} m_{2k-2}(f^\epsilon)(s) \, ds.$$

By induction we deduce that

$$m_{2k}(f^\epsilon)(t) \leq e^{-2k\beta t} m_{2k}(f_o^\epsilon) + 2k(2k-1)\sigma e^{-2k\beta t} A_{2k-2} \int_0^t e^{2k\beta s} \, ds.$$

Thus, we conclude the lemma. $\square$

Proposition 3.2 For any $\alpha > 1$ and any $t \geq 0$ we have

$$\|e^{\frac{\beta}{2\alpha\sigma}v^2} f^\epsilon(t)\|_{L^1(\mathbb{R})} \leq \frac{\alpha}{\alpha-1} \|e^{\frac{\beta}{2\sigma}v^2} f_o\|_{L^1(\mathbb{R})}.$$

and

$$\|(1+v^{2k})f^\epsilon(t)\|_{L^1(\mathbb{R})} \leq C \, \|(1+v^{2k}f_o\|_{L^1(\mathbb{R})}$$

for any k with C a constant depending on k. Moreover, we deduce that $H(f^\epsilon)$ is bounded independently of ϵ.

Proof .- Previous lemma implies by induction that

$$\frac{A_{2k}}{k!} \leq \sum_{i=0}^{k} \left(\frac{2\sigma}{\beta}\right)^i \frac{m_{2(k-i)}(f_o^\epsilon)}{(k-i)!}$$

for any $k \geq 0$, because

$$\frac{A_{2k}}{k!} \leq \frac{m_{2k}(f_o^\epsilon)}{k!} + \frac{2\sigma}{\beta}\frac{A_{2k-2}}{(k-1)!}$$

for any $k \geq 1$. Due to the estimate

$$\sum_{k=0}^{n} \left(\frac{\beta}{2\alpha\sigma}\right)^k \frac{m_{2k}(f^\epsilon)(t)}{k!} \leq \sum_{k=0}^{n} \left(\frac{\beta}{2\alpha\sigma}\right)^k \frac{A_{2k}}{k!}$$

we have that

$$\begin{aligned}\sum_{k=0}^{n} \left(\frac{\beta}{2\alpha\sigma}\right)^k \frac{m_{2k}(f^\epsilon)(t)}{k!} &\leq \sum_{k=0}^{n} \left(\frac{\beta}{2\alpha\sigma}\right)^k \left(\sum_{i=0}^{k} \left(\frac{2\sigma}{\beta}\right)^i \frac{m_{2(k-i)}(f_o^\epsilon)}{(k-i)!}\right) \\ &= \sum_{k=0}^{n} \alpha^{-k} \left(\sum_{i=0}^{k} \left(\frac{\beta}{2\alpha\sigma}\right)^{k-i} \frac{m_{2(k-i)}(f_o^\epsilon)}{(k-i)!}\right) \\ &\leq \|e^{\frac{\beta}{2\sigma}v^2} f_o\|_{L^1(\mathbb{R})} \sum_{k=0}^{n} \alpha^{-k}.\end{aligned}$$

As a consequence of Lemma 3.1 and taking into account that

$$\|e^{\frac{\beta}{2\sigma}v^2} f_o^\epsilon\|_{L^1(\mathbb{R})} = \lim_{n\to\infty} \sum_{k=0}^{n} \left(\frac{\beta}{2\sigma}\right)^k \frac{m_{2k}(f_o^\epsilon)}{k!}$$

we conclude the proof of the first estimate. The second estimate is obvious by induction.

The last estimate comes from the following bound in the evolution equation (3.6) for $H(f^\epsilon)$,

$$\frac{d}{dt}H(f^\epsilon) \leq \beta + 2\lambda \int_{\mathbb{R}^2} |\bar{v}-v| \, f^\epsilon \bar{f}^\epsilon \, dv \, d\bar{v}.$$

Since the right hand side can be estimated by the second moment of f, we deduce that $H(f^\epsilon)$ is bounded above in bounded time intervals independently of ϵ. To estimate it from below a standard argument involving the maxwellian tells you that

$$\int_{\mathbb{R}} f^\epsilon \, \log^- f^\epsilon \, dv$$

can be controled by the second moment and the mass of f^ϵ which are uniformly bounded. □

Since $\phi^\epsilon(v) \le |v|^2$ we have the estimate

$$\begin{aligned}\frac{|F^\epsilon(t,v)|}{1+v^2} &\le \frac{1}{1+v^2}\int_{\mathbb{R}} |\bar v - v|^2 f^\epsilon(t,x,\bar v)\, d\bar v \\ &\le \frac{1}{1+v^2}\left(m_2(f^\epsilon)(t) + v^2\|f_o\|_{L^1(\mathbb{R})}\right).\end{aligned}$$

Taking into account the previous corollary we have that

$$\frac{|F^\epsilon(t,v)|}{1+v^2}$$

is bounded in $L^\infty((0,\infty)\times\mathbb{R})$ independently of ϵ. Analogously, we have

$$\partial_v\left(\frac{F^\epsilon(t,v)}{1+v^2}\right) = \frac{\partial_v F^\epsilon(t,v)}{1+v^2} + \frac{-2vF^\epsilon(t,v)}{(1+v^2)^2}$$

and

$$|\partial_v F^\epsilon(t,v)| \le C(1+|v|)$$

and thus,

$$\partial_v\left(\frac{F^\epsilon(t,v)}{1+v^2}\right)$$

is bounded in $L^\infty((0,\infty)\times\mathbb{R})$ independently of ϵ. If we are able to prove equicontinuity in time, we will have strong convergence in $C([0,T]\times K)$ for any $T>0$ and any compact interval K. Let us take $t \ge s \ge 0$. It is easy to check that

$$\begin{aligned}&\frac{|F^\epsilon(t,v) - F^\epsilon(s,v)|}{1+v^2} \\ &\le \frac{2}{1+v^2}\left(\|v^2(f^\epsilon(t,v) - f^\epsilon(s,v))\|_{L^1(\mathbb{R})} + v^2\|f^\epsilon(t,v) - f^\epsilon(s,v)\|_{L^1(\mathbb{R})}\right).\end{aligned}$$

Similar estimates to (3.3) and (3.4) for the existence of the regularized equation imply that

$$\begin{aligned}\|f^\epsilon(t) - f^\epsilon(s)\|_{L^1(\mathbb{R})} &\le \lambda C\int_s^t e^{\beta s'}\, \nu(t-s')^{-\frac12}\, \|(1+v^2)^{-1}F^\epsilon(s')\|_{L^\infty(\mathbb{R})} \\ &\quad \|(1+v^2)f^\epsilon(s')\|_{L^1(\mathbb{R})}\, ds' \le C(\Phi(t)-\Phi(s)). \qquad (3.7)\end{aligned}$$

Similarly, we have

$$\begin{aligned}&\|v^2(f^\epsilon(t) - f^\epsilon(s))\|_{L^1(\mathbb{R})} \\ &\le \lambda C\int_s^t e^{\beta s'}\|v^2(-\partial_v M_{\nu(t-s')} * (F^\epsilon(s')f^\epsilon(s'))_{\alpha(s')^{-1}})_{\alpha(t)}\|_{L^1(\mathbb{R})}\, ds'.\end{aligned}$$

Using that $v^2 \le 2((v-\bar v)^2 + \bar v^2)$ and

$$|(v-\bar v)^2\partial_v M_{\nu(t-s')}(v-\bar v)| \le C\nu(t-s')|\partial_v^3 M_{\nu(t-s')}(v-\bar v)|$$

we can estimate it as

$$\|v^2(f^\epsilon(t) - f^\epsilon(s))\|_{L^1(\mathbb{R})} \le \lambda C\int_s^t e^{\beta s'}\, \nu(t-s')^{-\frac12}\, \|(1+v^2)F^\epsilon(s')f^\epsilon(s')\|_{L^1(\mathbb{R})}\, ds'.$$

Because of fourth order moments of f^ϵ are uniformly bouded in ϵ we finally have again

$$\|v^2(f^\epsilon(t) - f^\epsilon(s))\|_{L^1(\mathbb{R})} \le C(\Phi(t)-\Phi(s)).$$

Therefore, Ascoli-Arzela theorem gives us the strong convergence of

$$\frac{F^\epsilon(t,v)}{1+v^2}$$

in $C([0,T]\times K)$ for any $T>0$ and K compact interval.

Using the properties stated in corollary 3.2, it is easy to prove the weak L^1 convergence of the densities f^ϵ for any $t>0$. To apply the Ascoli-Arzela version for weak convergence (see [7]) one has to prove the equicontinuity in time of the dual action of f^ϵ. This is a trivial consequence of (3.7). Therefore, f^ϵ is weakly compact in $C_w([0,T],L^1(\mathbb{R}))$ for any $T>0$. The same arguments prove that $(1+v^2)f^\epsilon$ is weakly compact in $C_w([0,T],L^1(\mathbb{R}))$ because of the boundeness of the fourth moment.

Taking subsequences, we have that as $\epsilon_n\to 0$ we obtain two limit functions $F(t,v)$ and $f(t,v)$. Let us see that $F=(-\phi)*f$. This is a consequence of the estimate

$$|F^\epsilon-(-\phi)*f| \;\le\; \int_{\mathbb{R}} |\phi(v-\bar v)-(v-\bar v)Abs^\epsilon(v-\bar v)|f^\epsilon(t,\bar v)\,d\bar v$$

$$\int_{\mathbb{R}} |\mathcal{X}^\epsilon(v-\bar v)-1||\phi(v-\bar v)|f^\epsilon(t,\bar v)\,d\bar v \;+\; |\int_{\mathbb{R}} \phi(v-\bar v)(f^\epsilon(t,\bar v)-f(t,\bar v)\,d\bar v|,$$

the uniform convergence of $\phi-vAbs^\epsilon$ to zero and the weak convergence of $(1+v^2)f^\epsilon$.

Finally, let us remark that similar estimates as above prove that $v\partial_v f^\epsilon$ is bounded in L^1 independently of ϵ and therefore the convergence of f^ϵ is strong.

Collecting all the information above on the approximate solutions we can finally prove the existence and uniqueness result (Theorem 1.1) for the equation (1.5).

Proof .- The uniqueness can be proved by standard arguments (see [5]) using the estimates of the fundamental solution given in section 2. The existence of the solution and the properties 1 and 2 is just the summary of all previous arguments.

The solution verifies the integral equation stated in property 4 passing to the limit in the integral equation verified by f^ϵ. The regularity of the solution for smooth initial data comes from the regularity of general uniformly parabolic equations and a boot-strap argument (cf. [5]). □

4 Asymptotic behavior and hydrodynamics

The asymptotic behavior of the solutions of (1.5) was analyzed in [3]. Therefore, we will review here these results and we obtain the formal hydrodynamics for (1.3).

It is proved in [3] that any solution of (1.5) converges as $t\to\infty$, to an equilibrium which is not Maxwellian. Such stationary solution is described implicitly by an equation of mean-field type, and, for large v, behaves as exp $C|v|^3$. In other words, the inelastic interaction makes the equilibrium distribution more picked around zero with respect to the usual Maxwellian distribution.

The basic tool in the asymptotic analysis is the following free-energy functional:

$$\eta(f)=\int f(v)\log f(v)dv+\frac{\lambda}{6\sigma}\int |v-\bar v|^3 f(v)f(\bar v)dvd\bar v+\frac{\beta}{2\sigma}\int v^2 f(v)dv.$$

The first result is that the stationary solution is the unique minimizer of the functional η over the probability measures with zero mean. Precisely we have

Theorem 4.1 [3] *Consider the functional* η *defined by* (2.11) *on the set:*

$$\mathcal{P}=\left\{f\in L^1(\mathbb{R})|f\ge 0,\int f=1,\int vf(v)=0\right\}.$$

Then η has a unique minimum, $f_\infty \in \mathcal{P}$, $f_\infty \in C^\infty$ and satisfies

$$\lambda \partial_v (F_\infty f_\infty) \;=\; \beta \partial_v (v f_\infty) + \sigma \partial_v^2 f_\infty \quad \textit{in } \mathbb{R}.$$

A remarkable property of the free energy functional is that it decreases along the solutions. Indeed if $f = f(t,v)$ is a classical solution with $f_0 \in \mathcal{P} \cap C^2(\mathbb{R})$, and $\eta(f_0) < +\infty$, then $\eta(f(t)) \leq \eta(f_0)$. Indeed by a direct computation and an integration by parts:

$$\dot{\eta} = -\int \left(\log f(v) + \frac{\lambda}{3\sigma}\int |v-\bar{v}|^3 f(\bar{v}) d\bar{v} + \frac{\beta}{2\sigma} v^2\right) \partial_v \left((\lambda F - \beta v) f - \sigma \partial_v f\right) dv$$

$$= -\frac{1}{\sigma}\int \frac{1}{f} |(\lambda F - \beta v) f - \sigma \partial_v f|^2 dv \leq 0$$

The previous properties of η implies that η is a Lyapunov functional for this system. It can be proved that this implies that f_∞ is L^1 asymptotically stable for (1.5).

Theorem 4.2 [3] *For any $f_0 \in \mathcal{P} \cap C^2(\mathbb{R})$, with $\int v^4 f_0 < +\infty$, and with $\eta(f_0) < +\infty$ we have*

$$\lim_{t \to \infty} \|f(t) - f_\infty\|_{L^1} = 0.$$

It is not known whether this solution is a stable stationary solution for the general non homogeneous case. However, one can develop a formal hydrodynamics by a Hilbert expansion method about the homogeneous stationary solution f_∞ considered as stationary solution for the non-homogeneous equation (1.3). Let us do it in the case where $\beta = 0$. This procedure does not work for $\beta > 0$.

Choosing reference values for the position and the velocity x_o and v_o and taking as time reference value $t_0 = \frac{x_o}{v_o}$ we consider the dimensionless variables $t = t_o \tilde{t}$, $x = x_o \tilde{x}$ and $v = v_o \tilde{v}$ with $f(t,x,v) = \tilde{f}(\tilde{t},\tilde{x},\tilde{v})$. Computing the equation for $\tilde{f}$ and dropping the tildes, we find

$$\epsilon \partial_t f + \epsilon v \partial_x f^\epsilon + \tilde{\lambda} \partial_v (F f) - \partial_v^2 f \;=\; 0$$

where ϵ and $\tilde{\lambda}$ are computed in terms of λ, σ and the reference values, see [4] for details.

Considering now the limit $\epsilon \to 0$, the equation for f^ϵ reads

$$\epsilon \partial_t f^\epsilon + \epsilon v \partial_x f^\epsilon + \tilde{\lambda} \partial_v (F^\epsilon f^\epsilon) - \partial_v^2 f^\epsilon \;=\; 0 \tag{4.1}$$

for any $\epsilon > 0$. Let us define the particle density and the mean velocity by

$$\rho^\epsilon(t,x) = \int_{\mathbb{R}} f^\epsilon(t,x,v)\, dv$$

and

$$\rho^\epsilon u^\epsilon(t,x) = \int_{\mathbb{R}} v f^\epsilon(t,x,v)\, dv\ .$$

Integrating (4.1) over v we find the continuity equation

$$\frac{\partial \rho^\epsilon}{\partial t} + \frac{\partial \rho^\epsilon u^\epsilon}{\partial x} \;=\; 0. \tag{4.2}$$

Multiplying (4.1) by v and integrating over v we obtain

$$\frac{\partial \rho^\epsilon u^\epsilon}{\partial t} + \frac{\partial \rho^\epsilon (u^\epsilon)^2}{\partial x} \;=\; -\frac{\partial p^\epsilon}{\partial x} \tag{4.3}$$

where p^ϵ is the pressure given by the fluctuations of the velocity, that is,

$$p^\epsilon(t,x) = \int_{\mathbb{R}} (v - u^\epsilon(t,x))^2 f^\epsilon \, dv \ .$$

Formally, we look for the solution of (4.1) by expanding in power series of ϵ,

$$f = \sum_{k \geq 0} \epsilon^k \ f_k.$$

Here, F_k is the convolution of $-\phi$ with f_k. Then up to order ϵ^0 in equation (4.1) we find

$$\lambda \partial_v (F_0 f_0) - \sigma \partial_v^2 f_0 \ = \ 0$$

and thus, f_0 is a stationary homogeneous solution of (1.3). The general set of stationary homogeneous solutions of (1.3) is given by

$$f_0(t,x,v) = \rho(t,x)^{4/3} f_\infty(\rho(t,x)^{1/3}(v - u(t,x))),$$

for any functions ρ and u. This can be easily checked by scaling arguments. Let us see the equations that ρ and u must satisfy.

Matching terms of order ϵ in equation (4.1) we have

$$\partial_t f_0 + v \partial_x f_0 + \lambda \partial_v (F_0 f_1 + F_1 f_0) - \sigma \partial_v^2 f_1 \ = \ 0.$$

Computing the first two moments of the previous identity we finally find

$$\frac{\partial \rho}{\partial t} + \frac{\partial \rho u}{\partial x} \ = \ 0 \tag{4.4}$$

and

$$\frac{\partial \rho u}{\partial t} + \frac{\partial \rho u^2}{\partial x} \ = \ -\frac{\partial p}{\partial x} \tag{4.5}$$

where

$$p(t,x) = \int_{\mathbb{R}} (v - u(t,x))^2 f_o \, dv \ = \ A_\infty \rho^{1/3}$$

and A_∞ is a constant depending on f_∞. Also, this system appears substituting f^ϵ by f_0 in equations (4.2) and (4.3).

Therefore the system (4.4)-(4.5) is the hydrodynamical system using the Hilbert expansion for the scaled equation (4.1). More information and a deeper discussion about the hydrodynamics of this system can be found in [4].

References

[1] Benedetto, D. and Caglioti E., 'The Collapse Phenomenon in one dimensional inelastic point particle systems', to appear in *Physica D.*

[2] Benedetto, D., Caglioti E. and Pulvirenti, M., 'A kinetic equation for granular media', *Modél. Math. Anal. Numér.* **31**, 615-641 (1997).

[3] Benedetto, D., Caglioti E., Carrillo, J. A. and Pulvirenti, M., 'A non Maxwellian equilibrium distribution for one-dimensional granular media', *J. Stat. Phys.* **91**, 979-990 (1998).

[4] Benedetto, D., Caglioti E., Golse, F. and Pulvirenti, M., 'A hydrodynamic model arising in the context of granular media', to appear in *Computers and Mathematics with Applications.*

[5] Bouchut, F., 'Existence and Uniqueness of a Global Smooth Solution for the Vlasov-Poisson-Fokker-Planck System in Three Dimensions', *J. Func. Anal.* **111**, 239–258 (1993).

[6] Carrillo, J. A. 'Global weak solutions for the initial-boundary value problems to the Vlasov-Poisson-Fokker-Planck system', *Math. Meth. in the Appl. Sci.* **21**, 907-938 (1998).

[7] Carrillo, J. A. and Soler, J., 'On the initial value problem for the Vlasov-Poisson-Fokker-Planck system with initial data in L^p spaces', *Math. Meth. in the Appl. Sci.* **18**, 825-839 (1995).

[8] Carrillo, J. A., Toscani, G., 'Exponential convergence toward equilibrium for homogeneous Fokker-Planck-type equations', *Math. Meth. in the Appl. Sci.* **21**, 1269-1286 (1998).

[9] Cercignani, C., Illner, R. and Pulvirenti, M., *The mathematical theory of dilute gases*, Springer series in Applied Mathematical Sciences **106**, Springer-Verlag, New-York, 1994.

[10] Horst, E., 'On the classical solutions of the initial value problem for the unmodified nonlinear Vlasov Equations. I: General theory.', *Math. Meth. in the Appl. Sci.* **3**, 229-248 (1981).

[11] Rein, G. and Weckler, J., 'Generic global classical solutions of the Vlasov-Fokker-Planck-Poisson system in three dimensions', *J. Differential Equations* **99**, 59-77 (1992).

[12] Victory, H. D. and O'Dwyer, B. P., 'On classical solutions of Vlasov-Poisson-Fokker-Planck systems', *Indiana Univ. Math. J.* **39**, 105-157 (1990).

Fields Institute Communications
Volume **27**, 2000

A Class of Cellular Automata Equivalent to Deterministic Particle Systems

Henryk Fukś
The Fields Institute for Research
in Mathematical Sciences
Toronto, Ontario M5T 3J1, Canada
and
Department of Mathematics and Statistics,
University of Guelph,
Guelph, Ontario N1G 2W1, Canada
hfuks@fields.utoronto.ca

Abstract. We demonstrate that a local mapping f in a space of bisequences over $\{0,1\}$ which conserves the number of nonzero sites can be viewed as a deterministic particle system evolving according to a local mapping in a space of increasing bisequences over $\mathbb{Z}$. We present an algorithm for determination of the local mapping in the space of particle coordinates corresponding to the local mapping f.

1 Introduction

Cellular automata (CA) are dynamical systems characterized by discreteness in space and time. In general, they can be viewed as cells in a regular lattice updated synchronously according to a local interaction rule, where the state of each cell is restricted to a finite set of allowed values. Among many applications of CA, models of road traffic flow, first proposed by Nagel and Schreckenberg in 1992 [12], attracted substantial attention in recent years. Many theoretical aspect of the Nagel-Schreckenberg model are still not fully understood, and therefore several simplified models have been proposed, including models based on deterministic cellular automata [11, 6].

On of the interesting features of these models is the fact that they can be described using two equivalent paradigms: either as one-dimensional cellular automata or as systems of interacting particles on one-dimensional lattice. The simplest example is rule 184, one of the elementary CA rules investigated by Wolfram [15], and later extensively studied in the context of surface growth models [10], as

2000 *Mathematics Subject Classification.* Primary 68Q80; Secondary 37B15.

The author wishes to thank The Fields Institute for Research in Mathematical Sciences for generous hospitality and the Natural Sciences and Engineering Research Council of Canada for financial support in the form of a postdoctoral fellowship.

well as in the context of density classification problem [4]. It is one of the only two (symmetric) non-trivial elementary rules conserving the number of active sites [1, 2], and, therefore, can be interpreted as a rule governing dynamics of particles (cars). Particles (cars) move to the right if their right neighbor site is empty, and do not move if the right neighbor site is occupied, all of them moving simultaneously at each discrete time step. Using terminology of lattice stochastic processes, rule 184 can be viewed as a discrete-time version of totally asymmetric simple exclusion process.

A general question which can be asked is when a given CA rule can be treated as a rule governing motion of particles, and how to determine all rules possessing this property. Since the number of particles has to be conserved, it is clear that all such rules must conserve the number of nonzero sites. In [1, 2], CA rules of this type and their phenomenology were investigated.

In this work, we will formalize concepts introduced in [1]. We will prove that each conservative CA can be associated with a local mapping of particle coordinates, and we will demonstrate how such a mapping can be constructed.

2 Bisequence spaces and their mappings

The set of definitions given below closely follows terminology used in [9], with minor modifications.

Let $\mathcal{G}$ be a countable set with cardinality g, which will be called a *symbol set.* If g is finite, we will assume that $\mathcal{G} = \{0, 1, \dots, g-1\}$, otherwise we will often assume $\mathcal{G} = \mathbb{Z}$.

A *bisequence* over $\mathcal{G}$ is a function on $\mathbb{Z}$ to $\mathcal{G}$. Let $X(\mathcal{G})$ denote the set of bisequences over $\mathcal{G}$, i.e., $X(\mathcal{G}) = \mathcal{G}^{\mathbb{Z}}$. If $x \in X(\mathcal{G})$ and $i \in \mathbb{Z}$, then $x(i)$ will be often denoted by x_i.

Let $n \in \mathbb{N}, n > 0$. An *n-block* over $\mathcal{G}$ is an ordered set $x_1 x_2 \dots x_n$, where $x_i \in \mathcal{G}$. The set of all n-blocks over $\mathcal{G}$ will be denoted by $\mathcal{B}_n(\mathcal{G})$.

Let f be a mapping of $\mathcal{B}_n(\mathcal{G})$ into $\mathcal{G}$. The set of all such mappings, for a given $n \in \mathbb{N}, n > 0$ will be denoted by $F(\mathcal{G}, n)$. Mapping f will be frequently called a *local mapping*, or equivalently a *cellular automaton rule.*

Let $f \in F(\mathcal{G}, n)$ and let $m \in \mathbb{N}$, $m > 0$. Corresponding to f and m, we define a mapping $f_m : \mathcal{B}_{m+n-1}(\mathcal{G}) \to \mathcal{B}_m(\mathcal{G})$ as follows. Let $b = b_1 \dots b_{m+n-1} \in \mathcal{B}_{m+n-1}(\mathcal{G})$ and let $a_i = f(b_i b_{i+1} \dots b_{i+n-1})$, where $i = 1, 2, \dots, m$. Then we define $f_m(B) = a$, where $a = a_1 a_2, \dots, a_m$.

Finally, let us define a *global mapping* $f_\infty : X(\mathcal{G}) \to X(\mathcal{G})$ corresponding to a local mapping $f \in F(\mathcal{G}, n)$. Let $x \in X(\mathcal{G})$, and let $y \in X(\mathcal{G})$ be defined as $y_i = f(x_i, x_{i+1}, \dots, x_{i+n-1})$ for all $i \in \mathbb{Z}$. Then we define $f_\infty(x) = y$.

For $a \in \mathbb{Z}$ we can define a mapping $f_{\infty,a} : X(\mathcal{G}) \to X(\mathcal{G})$, similar to f_∞. If $x \in X(\mathcal{G})$, and if $y \in X(\mathcal{G})$ such that $y_i = f(x_{i-a}, x_{i+1-a}, \dots, x_{i+n-1-a})$ then, by definition, $f_{\infty,a} = y$. Note that $f_\infty = f_{\infty,0}$.

As an example, consider a set of mappings $F(\{0,1\}, 3)$, which were studied in [15] and called elementary cellular automata rules. There is $2^{2^3} = 256$ such mappings, and it is customary, following [15], to assign them *code numbers* $C(f)$ defined as

$$C(f) = \sum_{x_1, x_2, x_3 = 0}^{1} f(x_1, x_2, x_3) 2^{(2^2 x_1 + 2^1 x_2 + 2^0 x_3)}. \tag{2.1}$$

Example 2.1 For instance, the local mapping with code number 184, often referred to as simply *rule 184*, is defined by

$$\begin{aligned} &f(0,0,0) = 0,\ f(0,0,1) = 0,\ f(0,1,0) = 0,\ f(0,1,1) = 1 \\ &f(1,0,0) = 1,\ f(1,0,1) = 1,\ f(1,1,0) = 0,\ f(1,1,1) = 1. \end{aligned} \tag{2.2}$$

It is straightforward to show that the above definition can be written in a more compact form as

$$f(x_1, x_2, x_3) = x_2 + \min\{x_1, 1 - x_2\} - \min\{x_2, 1 - x_3\}. \tag{2.3}$$

Having the local mapping, we can compute m-block mappings f_m. For example, $f_3(10100) = 101$ because $f(1,0,1) = 1$, $f(0,1,0) = 0$, and $f(1,0,0) = 1$.

It is a common practice to view CA as dynamical systems and to investigate trajectories of points in the bisequence space $X(\mathcal{G})$, where by a trajectory of a point $x \in \mathcal{G}$ we mean the sequence $\{f_\infty^k(x)\}_{k=0}^\infty$, superscript k denoting multiple composition

$$f_\infty^k = \underbrace{f_\infty(f_\infty(\cdots f_\infty(x)))}_{k \text{ times}}. \tag{2.4}$$

Since CA are often used as model of real physical systems of a finite size, periodic boundary conditions are frequently employed. Using the formalism introduced in this section, periodic boundary conditions can be easily incorporated by assuming that the bisequence under consideration is periodic with some period L, i.e., there exist some $L \in \mathbb{N}$ such that $\forall i \in \mathbb{Z} : x_{i+L} = x_i$. Obviously, all bisequences belonging to the trajectory of a periodic bisequence are periodic.

3 Conservative rules

As in any other dynamical system, symmetry and conservation laws play an important role in cellular automata. Additive invariants in one-dimensional CA have been studied by Hattori and Takesue [8, 14]. They obtained conditions which guarantee the existence of additive conserved quantities, and produced a table of additive invariants for Wolfram's elementary CA rules. In this work, we will consider simplest possible additive invariants, namely the number of active sites ("active" meaning non-zero).

Let us start from a simple example, rule 184 introduced in the previous section. Let us consider an L-periodic bisequence $x \in X(\{0,1\})$ and its image under the local mapping with code number 184, as defined by eq. (2.3), $y = f_\infty(x)$. The number of active sites in a single period of x (to be referred to as simply the *number of active sites in x*) is equal to $\sum_{i=1}^L x_i$. The number of active sites in y is, using eq. (2.3),

$$\sum_{i=1}^{i=L} y_i = \sum_{i=1}^{i=L} f(x_{i+1}, x_{i+2}, x_{i+3}) = \sum_{i=1}^{i=L} x_{i+2} +$$

$$\sum_{i=1}^{i=L} \min\{x_{i+1}, 1 - x_{i+2}\} - \sum_{i=1}^{i=L} \min\{x_{i+2}, 1 - x_{i+3}\}.$$

Since for a periodic lattice the last two sums cancel, we obtain

$$\sum_{i=1}^{i=L} x_i = \sum_{i=1}^{i=L} y_i, \tag{3.1}$$

meaning that the number of active sites is conserved.

In general, a local mapping $f \in F(\mathcal{G}, n)$ will be called *L-conservative* if for any L-periodic bisequence $x \in X(\mathcal{G})$ the following condition is satisfied:

$$\sum_{i=1}^{i=L} f(x_i, x_{i+1}, \ldots, x_{i+n-1}) = \sum_{i=1}^{i=L} x_i. \tag{3.2}$$

Using the fact that x is periodic, this condition can be rewritten as

$$f(x_1, x_2, \ldots, x_n) + f(x_2, x_3, \ldots, x_{n+1}) + \cdots + f(x_L, x_1, \ldots, x_{n-1}) \\ = x_1 + x_2 + \cdots + x_L. \tag{3.3}$$

If for every $L > 0$ a mapping $f \in F(\mathcal{G}, n)$ is L-conservative, than it is said to be *conservative.* The following theorem [1] is helpful in determining if a given rule is conservative.

Theorem 3.1 ([1]) *If a mapping $f \in F(\mathcal{G}, n)$ is L-conservative for $L = 2n-2$, then it is conservative.*

Proof To prove the above result, we shall show that, if $L > 2n-2$, any equation of (3.3) type, i.e., equation expressing conservation condition for L-periodic configuration, is a linear combination of three equations which express, respectively, conservation conditions for $(L-1)$-, $(2n-3)$-, and $(2n-2)$-periodic sequences. More precisely, for all L-periodic bisequences with period L, $\{x_1, x_2, \ldots, x_L\}$, Equation (3.3) can be written

$$\Big(f(x_1, x_2, \ldots, x_n) + f(x_2, x_3, \ldots, x_{n+1}) + \cdots + f(x_{L-1}, x_1, \ldots, x_{n-1})\Big) \\ -\Big(f(x_1, x_2, \ldots, x_{n-2}, x_{L-n+1}, x_{L-n+2}) \\ +f(x_2, x_3, \ldots, x_{L-n+3}) + \cdots + f(x_{n-2}, x_{L-n+1}, \ldots, x_{L-1}) \\ +f(x_{L-n+1}, x_{L-n+2}, \ldots, x_{L-1}, x_1) + \ldots + f(x_{L-1}, x_1, \ldots, x_{n-2}, x_{L-n+1})\Big) \\ +\Big(f(x_1, x_2, \ldots, x_{n-2}, x_{L-n+1}, x_{L-n+2}) + f(x_2, x_3, \ldots, x_{L-n+3}) + \cdots \\ +f(x_{n-2}, x_{L-n+1}, \ldots, x_{L-1}) + f(x_{L-n+1}, x_{L-n+2}, \ldots, x_{L-1}, x_L) + \ldots \\ +f(x_{L-1}, x_L, x_1, \ldots, x_{n-2}) + f(x_L, x_1, \ldots, x_{n-2}, x_{L-n+1})\Big) \\ = (x_1 + x_2 + \cdots + x_{L-1}) - (x_1 + \cdots + x_{n-2} + x_{L-n+1} + \cdots + x_{L-1}) \\ +(x_1 + \cdots + x_{n-2} + x_{L-n+1} + \cdots + x_L). \tag{3.4}$$

The above rearrangement of terms[1] is valid if $x_{L-1} = x_L$, and for almost every periodic bisequence, we can choose a "coordinate system" (labeling of sites) in which it is true. The only exception is, for even L, the bisequence with period $1010\ldots10$. Verifying (3.4) is in this case possible, but needs to be done with a different method (see below). By induction, relation (3.4) shows that the conservation condition for L-periodic bisequence is a linear combination of conservation conditions for $(2n-3)$- and $(2n-2)$-periodic bisequences. Using similar rearrangement of terms as in (3.4), it can be shown that conservation conditions for $(2n-2)$-periodic bisequences imply conservation conditions for $(2n-2)$-periodic bisequences, concluding the proof.

[1]As pointed out by a referee, this rearrangement of terms can be also understood as a representation of an L-cycle in a de Bruijn graph as the sum of an $L-1$-cycle, a $2n-2$-cycle, and a $2n-3$-cycle.

Consider now the exception mentioned in the preceding paragraph, the bisequence 1010...10. Conservation condition for arbitrary L reads

$$\underbrace{f(1010\dots10) + f(0101\dots01) + \dots + f(0101\dots01)}_{L} = \frac{L}{2}, \tag{3.5}$$

if n is even, and

$$\underbrace{f(1010\dots01) + f(0101\dots10) + \dots + f(0101\dots10)}_{L} = \frac{L}{2}, \tag{3.6}$$

if n is odd. That is,

$$f(1010\dots10) + f(0101\dots01) = 1, \tag{3.7}$$

if n is even, and

$$f(1010\dots01) + f(0101\dots10) = 1, \tag{3.8}$$

if n is odd. In the above two equations L does not occur at all, so obviously, if eq. (3.5) or (3.6) is true for $L = 2n - 2$, it is true for any L. □

Using this theorem, if one wants to find all mappings $f \in F(\mathcal{G}, n)$ for some n which are conservative, it is enough to check which of them are $2n-2$-conservative. In fact, for many mappings it is possible to show that they are *not* conservative (and thus eliminate them from a list of mappings "suspected" for being conservative) by employing some of their elementary properties.

Theorem 3.2 *Let $n \in \mathbb{N}$ and $n > 0$. If a mapping $f \in F(\mathcal{G}, n)$ is conservative, then*

(a) $f(0, 0, \dots, 0) = 0$,
(b) $f(1, 1, \dots, 1) = 1$,
(c) $\displaystyle\sum_{x_1, x_2, \dots, x_n \in \{0,1\}} f(x_1, x_2, \dots, x_n) = 2^{n-1}$, *i.e., f is 1-balanced.*

Proof Part (a) becomes obvious if we consider periodic bisequence consisting of all zeros, i.e., containing no active sites. Image of this configuration must also contain no active sites, and it is only possible if $f(0, 0, \dots, 0) = 0$. Proof of (b) is identical.

Part (c) says that among all possible configurations of arguments $x_1, x_2, \dots, x_n$ of f (there are 2^n such configurations), exactly half results in $f(x_1, x_2, \dots, x_n) = 1$, as can be seen, for example, in definition of rule 184 (eq. 2.2).

To prove part (c), consider periodic bisequence $t \in X(\{0, 1\})$ of period $m = 2^n$. Now, let us construct a set of m blocks $A = \{b^{(j)}\}_{j=1}^{m}$ such that $b^{(j)} = t_j, t_{j+1}, \dots, t_{j+n-1}$ (superscript (j) denotes here just a consecutive number of the block $b^{(j)}$ in the set A).

Assume that we can find t such that all blocks in A are different. This means that each possible block of length n occurs in A once and only once, and therefore

$$\sum_{x_1, x_2, \dots, x_n \in \{0,1\}} f(x_1, x_2, \dots, x_n) = \sum_{i=1}^{m} f(t_i, t_{i+1}, \dots, t_{i+n-1}) = N_1, \tag{3.9}$$

where N_1 is the number of 1's in a single period of t.

On the other hand, $\bar{t} \in X(\{0,1\})$, which is obtained from t by replacing all zeros by ones and *vice versa*, must also have the same property as t, i.e., in a set $\bar{A} = \{b^{(j)}\}_{j=1}^m$ such that $b^{(i)} = \bar{t}_i, \bar{t}_{i+1}, \ldots, \bar{t}_{i+n-1}$, all elements are different, hence

$$\sum_{x_1,x_2,\ldots,x_n \in \{0,1\}} f(x_1,x_2,\ldots,x_n) = \sum_{i=1}^{m} f(\bar{t}_i, \bar{t}_{i+1}, \ldots, \bar{t}_{i+n-1}) = N_0, \tag{3.10}$$

where N_0 is the number of 0's in a single period of t. Comparing (3.9) and (3.10) we obtain $N_0 = N_1 = 2^n/2 = 2^{n-1}$, exactly as required.

The only problem left is to show that, indeed, for any $n > 0$, we can construct 2^n-periodic configuration t such that all blocks occurring in a single period of t are different (and therefore, constitute a set of all possible blocks of length n). For example, for $n = 2$, consider t with a period 1100. One can easily check that all possible blocks of length 2 occur in a single period of t: 11, 10, 00 and 01. For $n = 3$, t with a period 11101000 has the same property. Again, one can see that blocks occurring in the period of t, 111, 110, 101, 010, 100, 000, 001, and 011, are all possible blocks of length 3.

For a general n, the required t is equivalent to a hamiltonian cycle in the de Bruijn graph [3] of dimension n (or an eulerian cycle in the de Bruijn graph of dimension $n-1$). It has been demonstrated that such a cycle always exists (more precisely, for a given n, there exist exactly $2^{2^{n-1}-n}$ of such cycles – see, for example, review article [13]). □

4 Conservative cellular automata viewed as deterministic particle systems

Since conservative rules conserve the number of active sites, one can identify active sites with "particles" which change position after each application of the rule f_∞, but their number does not change. Therefore, instead of describing this system of particles in terms of lattice sites being occupied or empty, we can describe it by specifying list of coordinates of all particles.

Definition 4.1 An *increasing bisequence* over $\mathbb{Z}$ is an increasing function on $\mathbb{Z}$ to $\mathbb{Z}$. Set of all increasing sequences over $\mathbb{Z}$ will be denoted by $\tilde{X}(\mathbb{Z})$. If $s \in \tilde{X}(\mathbb{Z})$ and $i \in \mathbb{Z}$, then $s(i)$ will be often denoted by s_i. Now, let us define a mapping $\phi : \tilde{X}(\mathbb{Z}) \mapsto X(\{0,1\})$ as follows: let $s \in \tilde{X}(\mathbb{Z})$ be an increasing bisequence and let $x \in X(\{0,1\})$ be defined by

$$x_i = \begin{cases} 1, & \text{if } \exists k \in \mathbb{Z} : s_k = i, \\ 0, & \text{otherwise,} \end{cases} \tag{4.1}$$

for every $i \in \mathbb{Z}$. Then, by definition, $\phi(s) = x$.

Using the notion of particles, s in the above definition can be understood as a list of coordinates of particles on one-dimensional lattice. We require this list to be increasing, so no two particles occupy the same position. On the other hand, $x \in X(\{0,1\})$ is just a list of occupancy numbers: for all sites, x_i is either 1 (meaning site i is occupied) or 0 (empty site). Transformation ϕ takes a list of particle coordinates s and returns the corresponding list of occupancy numbers. For example, if in some $s \in \tilde{X}(\mathbb{Z})$ we have $s_1 = 7$, $s_2 = 9$, $s_3 = 12$, then in the corresponding $x = \phi(s)$, sites 7, 9, and 12 are occupied, meaning that $x_7 = 1$, $x_9 = 1$, and $x_{12} = 1$. Note that since s is an increasing bisequence, we can immediately conclude that the site between 7 and 9 as well as both sites between 9 and 12 must be empty, i.e., $x_8 = 0$ and $x_{10} = x_{11} = 0$.

Since conservative mappings conserve the number of particles, we would now like to find a local mapping which transforms coordinate lists $y \in \tilde{X}(\mathbb{Z})$ with the same effect as conservative local mappings $f \in (\{0,1\}), n)$ transform occupancy lists $x \in X(\{0,1\})$.

Theorem 4.2 *Let $n \in N$, and let $f \in F(\{0,1\}, n)$. There exist $m \in \mathbb{N}$, $a \in \mathbb{Z}$, and $g \in F(\mathbb{Z}, m)$ such that*

(i) *for every $s \in \tilde{X}(\mathbb{Z})$, $g_{\infty,a}(x) \in \tilde{X}(\mathbb{Z})$,*
(ii) *$\phi g_{\infty,a} = f_\infty \phi$.*

Note that Theorem 4.2 can be represented as

$$\begin{array}{ccc} \tilde{X}(\mathbb{Z}) & \xrightarrow{\ \phi\ } & X(\{0,1\}) \\ \big\downarrow{\scriptstyle g_{\infty,a}} & & \big\downarrow{\scriptstyle f_\infty} \\ \tilde{X}(\mathbb{Z}) & \xrightarrow{\ \phi\ } & X(\{0,1\}). \end{array} \tag{4.2}$$

Before we present the proof, let us consider an example which will clarify the meaning of Theorem 4.2.

Example 4.3 Equation 2.3 defined a local function for cellular automaton rule 184 as

$$f(x_1, x_2, x_3) = x_2 + \min\{x_1, 1 - x_2\} - \min\{x_2, 1 - x_3\}. \tag{4.3}$$

Corresponding rule g of Theorem 4.2 (constructed by an algorithm to be presented later) is defined as:

$$g(s_1, s_2) = s_1 + \min\{s_2 - s_1 - 1, 1\}. \tag{4.4}$$

Consider now a bisequence $s \in \tilde{X}(\mathbb{Z})$, in which $s_1 = 1$, $s_2 = 2$, and $s_3 = 4$, while $s_0 < 1$, $s_4 > 7$. (this means that among lattice sites $i = 1 \ldots 8$ only $i = 1, 2$ and 4 are occupied). Let $p = g_\infty(s)$, which means that $p_i = g(s_i, s_{i+1})$, yielding $s_1 = 1$, $s_2 = 3$, and $s_3 = 5$. List of particle coordinates $\{\ldots 1, 2, 4, \ldots\}$ is transformed by g_∞ into $\{\ldots 1, 3, 5, \ldots\}$

Now, consider $x \in X(\{0,1\})$ such that $x = \phi(s)$, and let us find $y = f_\infty(x)$, so that[2] $y_i = f(x_{i-1}, x_i, x_{i+1})$. Both x and y can be represented as

...	1	1	0	1	0	0	0	...
...	1	0	1	0	1	0	0	...

where the first line corresponds to x, and the second line to y. One can readily see that in the first line particles are located at sites 1, 2 and 4, while at second line at 1, 3 and 5, confirming that $\phi g_\infty = f_\infty \phi$. Note that in this example application of g_∞ results in a well defined motion of particles: if the site on the right of a given particle is empty, it moves to that site, otherwise it stays in the same place. Using notation introduced in [1], this can be written as

$$\overset{\curvearrowright}{10}, \quad \overset{\circlearrowleft}{11}. \tag{4.5}$$

[2]Strictly speaking, we are using $f_{\infty,1}$ here, not f_∞, mainly due to historical convention commonly employed for elementary (3-input) cellular automata. If we wanted to use $y_i = f(x_i, x_{i+1}, x_{i+2})$, we would have to slightly redefine g, so that $g(s_1, s_2) = s_1 - 1 + \min\{s_2 - s_1 - 1, 1\}$. However, the form used in the example is more intuitive (it represents asymmetric exclusion process with discrete time).

(arrow shows where the particle will move, circular arrow indicates that the particle stays in the same place).

5 Labeling scheme for lattice sites

In order to construct function g of Theorem 4.2, we will first introduce transformation of coordinate space $\tilde{X}(\mathbb{Z})$ to intermediate ("mixed") space Ψ, which combines both coordinates of particles and information about occupancy of lattice sites. This can be done by labeling occupied lattice sites with consecutive integers, and empty lattice sites with extra symbol $\bullet$, so that configurations of particles become bisequences over $\mathbb{Z} \cup \{\bullet\}$ (the extra symbol "$\bullet$" had to be introduced since "0" can be a particle label, and therefore cannot be used to denote empty sites).

Definition 5.1 Let $x \in \tilde{X}(\mathbb{Z})$. Let $y \in X(\mathbb{Z} \cup \{\bullet\})$ be a bisequence constructed as follows: $y_i = j$ if there exists $j \in \mathbb{Z}$ such that $x_j = i$, otherwise $y_i = \bullet$. By definition, $\psi(x) = y$. Bisequences y as defined above will be called *increasing labeling bisequences*, and the set of all such bisequences will be denoted by $\Psi = \psi(\tilde{X}(\mathbb{Z}))$.

Note that ψ transforms a bisequence of particle coordinates, such as $x = \ldots 1, 2, 4, \ldots$ into a bisequence $\psi(x)$:

⋯	1	2	•	3	•	•	•	⋯

which resembles $\phi(x)$, except that occupied sites are now labeled with unique integers, increasing from left to right (the label of an occupied site is always larger by 1 than the label of the closest occupied site on the left). If we define, for $k \in \mathbb{Z} \cup \{\bullet\}$,

$$\|k\| = \begin{cases} 1, & \text{if } k \in \mathbb{Z}, \\ 0, & \text{if } k = \bullet, \end{cases} \tag{5.1}$$

then, obviously, $(\phi(x))_i = \|(\psi(x))_i\|$ for every $i \in \mathbb{Z}$.

We will now define a mapping which plays a similar role in Ψ as f plays in $X(\{0,1\})$. Corresponding to f, mapping $\hat{f} \in F(\mathbb{Z} \cup \{\bullet\}, n)$ will be defined as follows.

Define

$$\begin{aligned} G(x_1, x_2, \ldots, x_n) = f(0, 0, \ldots, 0, \|x_1\|) + f(0, 0, \ldots, \|x_1\|, \|x_2\|) + \\ \ldots + f(0, \|x_1\|, \|x_2\|, \ldots, \|x_n\|) + f(\|x_1\|, \|x_2\|, \ldots, \|x_n\|), \end{aligned} \tag{5.2}$$

and

$$H(x_1, x_2, \ldots, x_n, k) = \min\{i \in \mathbb{N} : \sum_{j=1}^{i} \|x_j\| = k\}. \tag{5.3}$$

Now, function $\hat{f}$ is defined as

$$\hat{f}(x_1, x_2, \ldots, x_n) = \begin{cases} x_l, & \text{if } f(\|x_1\|, \|x_2\|, \ldots, \|x_n\|) = 1, \\ \bullet, & \text{otherwise}, \end{cases} \tag{5.4}$$

where $l = H(x_1, x_2, \ldots, x_n, G(x_1, x_2, \ldots, x_n))$.

Example 5.2 For rule 184, as defined in (2.3), applying the above definition we obtain for every $x_1, x_2, x_3 \in \mathbb{Z}$,

$$\hat{f}(\bullet, \bullet, \bullet) = \bullet, \hat{f}(\bullet, \bullet, x_1) = \bullet, \hat{f}(\bullet, x_1, \bullet) = \bullet, \hat{f}(\bullet, x_1, x_2) = x_1,$$

$$\hat{f}(x_1, \bullet, \bullet) = x_1, \hat{f}(x_1, \bullet, x_2) = x_1, \hat{f}(x_1, x_2, \bullet) = \bullet, \hat{f}(x_1, x_2, x_3) = x_2.$$

An immediate consequences of the definition of $\hat{f}$ is $\|\hat{f}(x_1, x_2, \ldots, x_n)\| = f(\|x_1,\|, \|x_2,\|, \ldots, \|x_n,\|)$, and therefore:

Lemma 5.3 *Let $n \in \mathbb{N}$, $n > 0$, and let $f \in F(\{0,1\}, n)$. Then for every $x \in \Psi$, $\|\hat{f}_\infty(x)\| = f_\infty(\|x\|)$.*

A crucial property of $\hat{f}$ is that an image of an increasing labeling bisequence under $\hat{f}_\infty$ is also an increasing labeling bisequence:

Theorem 5.4 *Let $n \in \mathbb{N}$, $n > 0$, and let $f \in F(\{0,1\}, n)$. If $x \in \Psi$, then $\hat{f}_\infty(x) \in \Psi$.*

Proof Let us consider an increasing labeling bisequence x and its image $y = \hat{f}_\infty(x)$. Consider two sites in y, y_i and y_j, $i < j$, such that all sites between them are empty, i.e. $y_k = \bullet$ for all $i < k < j$. We want to show that $y_j = y_i + 1$, since this would demonstrate that y is indeed an increasing labeling bisequence. Obviously, $y_i = \hat{f}(x_i, x_{i+1}, \ldots, x_{i+n-1})$ and $y_j = \hat{f}(x_j, x_{j+1}, \ldots, x_{j+n-1})$. There must be some p, q such that $y_i = x_p$ and $y_j = x_q$, which, according to the definition of $\hat{f}$, must satisfy

$$\begin{aligned} p =& i - 1 + H(x_i, x_{i+1}, \ldots, x_{i+n-1}, G(x_i, x_{i+1}, \ldots, x_{i+n-1})), \\ q =& j - 1 + H(x_j, x_{j+1}, \ldots, x_{j+n-1}, G(x_j, x_{j+1}, \ldots, x_{j+n-1})). \end{aligned} \tag{5.5}$$

As a first step of the proof, we will find relationship between $G(x_i, x_{i+1}, \ldots, x_{i+n-1}))$ and $G(x_j, x_{j+1}, \ldots, x_{j+n-1}))$.

Let us denote the number of nonzero sites in the block $x_i, x_{i+1}, \ldots, x_{j-1}$ by N, i.e., $N = \|x_i\| + \|x_{i+1}\| +, \ldots, +\|x_{j-1}\|$. Consider now a bisequence $t \in \Psi$ such that $t_l = x_l$ for $i \leq l \leq j + n - 1$, and $t_l = \bullet$ otherwise, and another one, $u \in \Psi$, such that $u_l = x_l$ for $j \leq l \leq j + n - 1$, and $t_l = \bullet$ otherwise. Due to the fact that f is conservative, we have

$$\sum_{l=-\infty}^{\infty} \|\hat{f}(t_l, t_{l+1}, \ldots, t_{l+n-1})\| = N + \sum_{l=-\infty}^{\infty} \|\hat{f}(u_l, u_{l+1}, \ldots, u_{l+n-1})\|. \tag{5.6}$$

Using definitions of t and u, and the fact that $f(0, 0, \ldots, 0) = 0$, equation (5.6) becomes

$$\begin{aligned} &\|\hat{f}(\bullet, \ldots, \bullet, x_i)\| + \|\hat{f}(\bullet, \ldots, \bullet, x_i, x_{i+1})\| + \ldots + \|\hat{f}(x_i, x_{i+1}, \ldots, x_{i+n-1})\| + \\ &\quad \|\hat{f}(x_{i+1}, x_{i+2}, \ldots, x_{i+n})\| + \ldots + \|\hat{f}(x_j, x_{j+1}, \ldots, x_{j+n-1})\| + \\ &\quad \|\hat{f}(x_{j+1}, x_{j+2}, \ldots, \bullet)\| + \ldots + \|\hat{f}(x_{j+n-1}, \bullet, \ldots, \bullet)\| - N + \\ &\|\hat{f}(\bullet, \ldots, \bullet, x_j)\| + \|\hat{f}(\bullet, \ldots, \bullet, x_j, x_{j+1})\| + \ldots + \|\hat{f}(x_j, x_{j+1}, \ldots, x_{j+n-1})\| + \\ &\quad \|\hat{f}(x_{j+1}, x_{j+2}, \ldots, \bullet)\| + \ldots + \|\hat{f}(x_{j+n-1}, \bullet, \ldots, \bullet)\|. \end{aligned} \tag{5.7}$$

However, we know that there are no particles between y_i and y_j, therefore

$$\|\hat{f}(x_l, x_{l+1}, \ldots, x_{l+n-1})\| = 0 \tag{5.8}$$

for all l such that $i < l < j$. Taking this into account, and using definition of G (eq. 5.2) we obtain, after some cancellations,

$$\begin{aligned} G(x_i, x_{i+1}, \ldots, x_{i+n-1}) + \|f(x_j, x_{j+1}, \ldots, x_{j+n-1})\| = \\ N + G(x_j, x_{j+1}, \ldots, x_{j+n-1}). \end{aligned} \tag{5.9}$$

Since $\|f(x_j, x_{j+1}, \ldots, x_{j+n-1})\| = 1$, this finally becomes

$$G(x_i, x_{i+1}, \ldots, x_{i+n-1}) = N - 1 + G(x_j, x_{j+1}, \ldots, x_{j+n-1}). \tag{5.10}$$

Having this relationship, note that, according to definition of H (eq. 5.3), x_p is k_p-th particle in the block $x_i, x_{i+1}, \ldots, x_{i+n-1}$ starting from the left, where $k_p = G(x_i, x_{i+1}, \ldots, x_{i+n-1})$. Obviously, it is also is k_p-th particle in the block $x_i, x_{i+1}, \ldots, x_{j+n-1}$ starting from the left.

Similarly, x_q is k_q-th particle in the block $x_j, x_{j+1}, \ldots, x_{j+n-1}$, where $k_q = G(x_j, x_{j+1}, \ldots, x_{j+n-1})$. Knowing that sites $x_i, \ldots, x_{j-1}$ contain exactly N particles, we conclude that x_q must also be $(N + k_q)$-th particle in the block $x_i, x_{i+1}, \ldots, x_{j+n-1}$. However, $k_q = k_p - N + 1$, therefore:

Particle x_p is k_p-th particle from the left in the block $x_i, x_{i+1}, \ldots, x_{j+n-1}$, while x_q is $(k_p + 1)$-th particle in the same block.

This means that there is no other particle between x_p and x_q, hence $x_q = x_p + 1$ (we assumed that x is an increasing labeling bisequence), and, finally, $y_j = y_i + 1$, which is exactly what what we wanted to show. □

6 Construction of local mapping in $\tilde{X}(\mathbb{Z})$

We are now ready to see how g of Theorem 4.2 can be constructed.

First, let us extend definition of the mapping ψ (Definition 5.1) to blocks of particle coordinates. Let $b \in \mathcal{B}_m(\mathbb{Z})$ be called an *increasing block* if $b_1 < b_2 < \ldots < b_m$. Set of all such blocks will be denoted by $\tilde{\mathcal{B}}_m(\mathbb{Z})$. For $m > 1$, $b \in \tilde{\mathcal{B}}_m(\mathbb{Z})$ let us define $c = \psi(b)$ such that $c \in \mathcal{B}_{b_m - b_1 + 1}(\mathbb{Z} \cup \{\bullet\})$ and that, for every $l \in \mathbb{N}$ satisfying $1 \le l \le b_m - b_1 + 1$, if there exist $k \in \mathbb{N}$ such that $b_k = l$, then $c_l = k - b_1 + 1$, otherwise $c_l = \bullet$. Mapping ψ, as before, transforms finite and increasing list of particle coordinates into finite block in particles are located at sites $b_l - b_1 + 1$ and labeled with labels l. For example, if $b = \{9, 10, 14\}$, then $\psi(b) = \{1, 2, \bullet, \bullet, 3\}$.

Similarly as we defined increasing labeling bisequences, we define *blocks with increasing labels* as elements of the set $\psi(\tilde{\mathcal{B}}_m(\mathbb{Z}))$. This set will be denoted by $\tilde{\mathcal{B}}_m(\mathbb{Z} \cup \{\bullet\})$

In Theorem 5.4, we proved that $\hat{f}_\infty$ maps increasing labeling bisequences to increasing labeling bisequences. The same is true for blocks with increasing labels (and can be proved using almost identical reasoning as in the proof of Theorem 5.4, therefore we omit the proof):

Corollary 6.1 *For any $m, n \in N$, $m \ge n$, and for $f \in F(\{0,1\}, n)$, if $b \in \tilde{\mathcal{B}}_m(\mathbb{Z} \cup \{\bullet\})$, then $\hat{f}_m(b) \in \tilde{\mathcal{B}}_{m-n+1}(\mathbb{Z} \cup \{\bullet\})$.*

Let us now assume that we have a list of particle coordinates $s \in \tilde{X}(\mathbb{Z})$, and we want to find new list of particles' positions t such that $\phi(t) = f_\infty(\phi(s))$, where $f \in F(\{0,1\}, n)$ for some $n \in \mathbb{N}$, $n > 1$. Using $\hat{f}$, this can be done as follows:

Algorithm 6.2 For each particle coordinate s_i perform the following steps:

1. construct a block $b = \{s_{i-n+1}, s_{i-n+2}, \ldots, s_{i+n-1}\}$
2. find block $c = \hat{f}_{s_{i+n-1} - s_{i-n+1} + 1}(\psi(b))$, this will be a block with increasing labels, with particles in their new positions
3. find $d = \psi^{-1}(c)$, this will be a block containing new positions of particles
4. find new coordinate of particle i, which is simply equal to d_i

Note that the size of the block b is chosen large enough to ensure that particle with label i remains in the block after step 2. For many rules, smaller neighborhood will suffice.

The above algorithm provides a way to compute t_i given s_i and coordinates of particles in the neighborhood of s_i. This means that it defines a local mapping g requested by theorem Theorem 4.2. In fact, it is possible to write explicit expression for g based on the Algorithm 6.2. Corresponding to $f \in F(\{0,1\},n)$, mapping $g : \mathcal{B}_{2n-1}(\mathbb{Z}) \to \mathbb{Z}$ is defined as follows:

$$g(b) = (\psi^{-1}(\hat{f}(\psi(b))))_n, \tag{6.1}$$

and parameter a of Theorem 4.2 equals to n.

As a final note, let us remark that the relationship of f, $\hat{f}$, and g can be represented as

$$\begin{array}{ccccc} \tilde{X}(\mathbb{Z}) & \xrightarrow{\psi} & \Psi & \xrightarrow{\|\cdot\|} & X(\{0,1\}) \\ \Big\downarrow g_{\infty,a} & & \Big\downarrow \hat{f}_\infty & & \Big\downarrow f_\infty \\ \tilde{X}(\mathbb{Z}) & \xrightarrow{\psi} & \Psi & \xrightarrow{\|\cdot\|} & X(\{0,1\}). \end{array} \tag{6.2}$$

We should also mention that g constructed here is not the only local mapping satisfying Theorem 5.4. Infinite number of local mappings with the same properties as g can be constructed as a superposition of g and the shift map on $\tilde{X}(\mathbb{Z})$ (by the shift map we mean $\sigma : \tilde{X}(\mathbb{Z}) \to \tilde{X}(\mathbb{Z})$ such that $(\sigma(x))_i = x_{i-1}$ for every $x \in \tilde{X}(\mathbb{Z})$ and every $i \in \mathbb{Z}$). All mappings $\sigma^k g$ as well as $\sigma^{-k} g$ for any $k \in \mathbb{N}$ could be used in place of g. As a convention, we will always use a local mapping such that if $x_i = i$ for every $i \in \mathbb{Z}$, and $y = g_\infty(x)$, then $y_i = i$, i.e., if all lattice sites are occupied in x, then in y they are in the same position.

7 Example: four-input conservative rules

Among 2^{2^4} local mappings $f \in F(\{0,1\},4)$ (four-input rules), only 22 rules are conservative [1]. For all of them, we computed $\hat{f}$, as shown in Table 1.

The first column in this table represents code number $C(f)$, defined similarly as (2.1):

$$C(f) = \sum_{x_1,x_2,x_3,x_4=0}^{1} f(x_1,x_2,x_3,x_4) 2^{(2^3 x_1 + 2^2 x_2 + 2^1 x_3 + 2^0 x_4)}. \tag{7.1}$$

The second column is a binary representation of $C(f)$, meaning that it is a sequence of 16 binary digits $a_{15}a_{14}\ldots a_0$ such that for every $x_1, x_2, x_3, x_4 \in \{0,1\}$

$$a_{2^3 x_1 + 2^2 x_2 + 2^1 x_3 + 2^0 x_4} = f(x_1,x_2,x_3,x_4). \tag{7.2}$$

Entries in the third column are constructed in a similar way. They are sequences $b_{15}b_{14}\ldots b_0$ such that if in the definition of $\hat{f}$ we have $\hat{f}(x_1,x_2,x_3,x_4) = x_j$, then we define

$$b_{2^3\|x_1\| + 2^2\|x_2\| + 2^1\|x_3\| + 2^0\|x_4\|} = x_j, \tag{7.3}$$

and if

$$\hat{f}(x_1,x_2,x_3,x_4) = \bullet, \tag{7.4}$$

then

$$b_{2^3\|x_1\| + 2^2\|x_2\| + 2^1\|x_3\| + 2^0\|x_4\|} = \bullet. \tag{7.5}$$

Table 1 Four-input rules conserving density of nonzero sites.

$C(f)$	Binary form	$\hat{f}$
43690	1010101010101010	$x_4 \bullet x_4 \bullet x_4 \bullet x_4 \bullet x_4 \bullet x_4 \bullet x_4 \bullet x_4 \bullet$
43944	1010101110101000	$x_3 \bullet x_2 \bullet x_3 \bullet x_1 x_1 x_3 \bullet x_2 \bullet x_3 \bullet \bullet\bullet$
47288	1011100010111000	$x_3 \bullet x_2 x_2 x_3 \bullet \bullet \bullet x_3 \bullet x_2 x_2 x_3 \bullet \bullet\bullet$
48268	1011110010001100	$x_3 \bullet x_2 x_2 x_3 x_3 \bullet \bullet x_3 \bullet \bullet \bullet x_3 x_3 \bullet \bullet$
48770	1011111010000010	$x_3 \bullet x_2 x_2 x_3 x_3 x_4 \bullet x_3 \bullet \bullet \bullet \bullet \bullet x_4 \bullet$
49024	1011111110000000	$x_2 \bullet x_1 x_1 x_1 x_1 x_1 x_1 x_2 \bullet \bullet \bullet \bullet \bullet \bullet\bullet$
51448	1100100011111000	$x_3 x_3 \bullet \bullet x_3 \bullet \bullet \bullet x_3 x_3 x_2 x_2 x_3 \bullet \bullet\bullet$
52428	1100110011001100	$x_3 x_3 \bullet \bullet x_3 x_3 \bullet \bullet x_3 x_3 \bullet \bullet x_3 x_3 \bullet \bullet$
52930	1100111011000010	$x_3 x_3 \bullet \bullet x_3 x_3 x_4 \bullet x_3 x_3 \bullet \bullet \bullet \bullet x_4 \bullet$
53184	1100111111000000	$x_2 x_2 \bullet \bullet x_1 x_1 x_1 x_1 x_2 x_2 \bullet \bullet \bullet \bullet \bullet \bullet$
56528	1101110011010000	$x_2 x_2 \bullet x_2 x_1 x_1 \bullet \bullet x_2 x_2 \bullet x_2 \bullet \bullet \bullet \bullet$
57580	1110000011101100	$x_3 x_3 x_4 \bullet \bullet \bullet \bullet \bullet x_3 x_3 x_4 \bullet x_3 x_3 \bullet \bullet$
58082	1110001011100010	$x_3 x_3 x_4 \bullet \bullet \bullet x_4 \bullet x_3 x_3 x_4 \bullet \bullet \bullet x_4 \bullet$
58336	1110001111100000	$x_2 x_2 x_2 \bullet \bullet \bullet x_1 x_1 x_2 x_2 x_2 \bullet \bullet \bullet \bullet\bullet$
59946	1110101000101010	$x_3 x_3 x_4 \bullet x_4 \bullet x_4 \bullet \bullet \bullet x_4 \bullet x_4 \bullet x_4 \bullet$
60200	1110101100101000	$x_2 x_2 x_2 \bullet x_3 \bullet x_1 x_1 \bullet \bullet x_2 \bullet x_3 \bullet \bullet\bullet$
61680	1111000011110000	$x_2 x_2 x_2 x_2 \bullet \bullet \bullet \bullet x_2 x_2 x_2 x_2 \bullet \bullet \bullet \bullet$
62660	1111010011000100	$x_2 x_2 x_2 x_2 \bullet x_3 \bullet \bullet x_2 x_2 \bullet \bullet \bullet x_3 \bullet \bullet$
63544	1111100000111000	$x_2 x_2 x_2 x_2 x_3 \bullet \bullet \bullet \bullet \bullet x_2 x_2 x_3 \bullet \bullet\bullet$
64524	1111110000001100	$x_2 x_2 x_2 x_2 x_3 x_3 \bullet \bullet \bullet \bullet \bullet \bullet x_3 x_3 \bullet \bullet$
65026	1111111000000010	$x_2 x_2 x_2 x_2 x_3 x_3 x_4 \bullet \bullet \bullet \bullet \bullet \bullet \bullet x_4 \bullet$
65280	1111111100000000	$x_1 x_1 x_1 x_1 x_1 x_1 x_1 x_1 \bullet \bullet \bullet \bullet \bullet \bullet \bullet \bullet$

This will become clear when we consider, for example, rule 53184 from Table 1. For this rule, in the third column we have $x_2 x_2 \bullet \bullet x_1 x_1 x_1 x_1 x_2 x_2 \bullet \bullet \bullet \bullet \bullet \bullet$. This means that for $x_1, x_2, x_3, x_4 \in \mathbb{Z}$

$$
\begin{aligned}
\hat{f}(x_1, x_2, x_3, x_4) &= \bullet \quad \text{if} \quad \|x_1, x_2, x_3, x_4\| = 0000 \\
\hat{f}(x_1, x_2, x_3, x_4) &= \bullet \quad \text{if} \quad \|x_1, x_2, x_3, x_4\| = 0001 \\
&\dots \\
\hat{f}(x_1, x_2, x_3, x_4) &= x_2 \quad \text{if} \quad \|x_1, x_2, x_3, x_4\| = 0111 \\
\hat{f}(x_1, x_2, x_3, x_4) &= x_1 \quad \text{if} \quad \|x_1, x_2, x_3, x_4\| = 1000 \\
&\dots \\
\hat{f}(x_1, x_2, x_3, x_4) &= x_2 \quad \text{if} \quad \|x_1, x_2, x_3, x_4\| = 1111,
\end{aligned}
$$

which is a set of 16 equations fully defining $\hat{f}$ for rule 53184. Column 3 of Table 1 represents sets of similar 16 equations in a condensed form for all conservative 4-input rules.

In order to reduce the number of rules for further considerations, we can exploit the fact that cellular automata rules obtained by spatial reflection or conjugation (interchanging zeros and ones in rule table), or both reflection and conjugation of a given rule have similar properties as the original rule. Let, for a given $f \in$

$F(\{0,1\}, n)$,

$$f_R(x_1, x_2, x_3, x_4) = f(x_4, x_3, x_2, x_1), \tag{7.6}$$

$$f_C(x_1, x_2, x_3, x_4) = 1 - f(1 - x_1, 1 - x_2, 1 - x_3, 1 - x_4), \tag{7.7}$$

$$f_{RC}(x_1, x_2, x_3, x_4) = f_C(x_4, x_3, x_2, x_1). \tag{7.8}$$

All rules in the set $\{f, f_R, f_C, f_{RC}\}$ have similar dynamics, therefore it is enough to consider only one of them. For conservative 4-input rules, we have seven such sets, or equivalence classes: {49024, 59946, 65026, 43944}, {53184, 58082, 64524, 47288}, {56528, 57580, 62660, 51448}, {58336, 52930, 63544, 48268}, {60200, 48770, 60200, 48770}, {61680, 52428, 61680, 52428}, and {65280, 43690, 65280, 43690}. We will now present mappings g for the first rule in each set, defined in terms of the step function Θ (for $m \in \mathbb{Z}$, $\Theta(m) = 1$ if $m > 0$, otherwise $\Theta(m) = 0$) and the delta function (for $m, n \in \mathbb{Z}$, $\delta_{m,n} = 1$ if $m = 1$, otherwise $\delta_{m,n} = 0$):

Rule 49024: $g(x_0, x1, x_2, x_3) = x_1 - 1 + \Theta(x_3 + x_2 - 2x_1 - 3)$,

Rule 53184: $g(x_0, x1, x_2, x_3) = x_1 - 1 + \Theta(x_2 - x_1 - 1)$,

Rule 56528: $g(x_0, x1, x_2, x_3) = x_1 - 1 + \delta_{2,x_2-x_1}$,

Rule 58336: $g(x_0, x1, x_2, x_3) = x_1 - 1 + \Theta(x_2 - x_1 - 2)$,

Rule 60200: $g(x_0, x1, x_2, x_3) = x_1 - 1 + \Theta(x_2 - x_1 - 2) - \delta_{1,x_2-x_1}\Theta(x_1 - x_0 - 1)$,

Rule 61680: $g(x_0, x1, x_2, x_3) = x_1 - 1$,

Rule 65280: $g(x_0, x1, x_2, x_3) = x_1$.

Note that for all the above rules, g is a function of only 4 arguments (with the exception of rule 60200, it is really a function of 3 arguments x_1, x_2, x_3, but for the sake of uniformity we always used x_0, x_1, x_2, x_3). For all these rules, $\phi g_{\infty,1} = \phi f_\infty$. Note that we could redefine g's and introduce $\tilde{g}(x_0, x_1, x_2, x_3) = g(x_0, x_1, x_2, x_3) + c$, where $c \in \mathbb{Z}$, and then we would have $\phi \tilde{g}_{\infty,1} = \phi f_{\infty,c}$.

8 Concluding remarks

We have demonstrated that a local mapping f in a space of bisequences over $\{0,1\}$ which conserve the number of nonzero sites can be viewed as a deterministic particle system evolving according to a local mapping g in a space of increasing bisequences over $\mathbb{Z}$. We also presented an algorithm for determination of the local mapping in the space of particle coordinates corresponding to mapping f.

Viewing conservative CA as systems of interacting particles is often very useful in solving problems related to CA dynamics, in particular problems which can be broadly characterized as *forward problems* [7]: given a CA rule, determine (predict) its properties. For instance, it is often easier to characterize the trajectory $\{g^i_\infty(s)\}_{i=0}^\infty$ of a point $s \in \tilde{X}(\mathbb{Z})$, then the trajectory $\{f^i_\infty(x)\}_{i=0}^\infty$ of a point $x \in X(\{0,1\}$. An example of such approach can be found in [5], where we considered properties of trajectories of configurations in simplified deterministic models of road traffic flow.

References

[1] N. Boccara and H. Fukś, *Cellular automaton rules conserving the number of active sites*, J. Phys. A: Math. Gen. **31** (1998), 6007–6018.

[2] N. Boccara and H. Fukś, *Number-conserving cellular automata*, submitted to Fundamenta Informaticae.

[3] N. G. de Bruijn, *A combinatorial problem*, Nederl. Akad. Wetensch. Proc. **49** (1946), 758–746.
[4] H. Fukś, *Solution of the density classification problem with two cellular automata rules*, Phys. Rev. E **55** (1997), 2081R–2084R.
[5] H. Fukś, *Exact results for deterministic cellular automata traffic models*, Phys. Rev. E **60** (1999), 197–202.
[6] M. Fukui and Y. Ishibashi, *Traffic flow in 1D cellular automaton model including cars moving with high speed*, J. Phys. Soc. Japan **65** (1996), 1868–1870.
[7] H. Gutowitz (ed.), *Cellular automata: Theory and experiment*, MIT Press/Bradford Books, Cambridge, Mass., 1991.
[8] T. Hattori and S. Takesue, *Additive conserved quantities in discrete-time lattice dynamical systems*, Physica D **49** (1991), 295–322.
[9] G. A. Hedlund, *Endomorphism and automorphism of the shift dynamical system*, Mathematical Systems Theory **3** (1968), 320–375.
[10] J. Krug and H. Spohn, *Universality classes for deterministic surface growth*, Phys. Rev. A **38** (1988), 4271–4283.
[11] K. Nagel and H. J. Herrmann, *Deterministic models for traffic jams*, Physica A **199** (1993), 254–269.
[12] K. Nagel and M. Schreckenberg, *A cellular automaton model for freeway traffic*, J. Physique I **2** (1992), 2221–2229.
[13] A. Ralston, *De Bruijn sequences – a model example of the interaction of discrete mathematics and computer science*, Math. Mag. **55** (1982), 131–143.
[14] S. Takesue, *Staggered invariants in cellular automata*, Complex Systems **9** (1995), 149–168.
[15] S. Wolfram, *Cellular automata and complexity: Collected papers*, Addison-Wesley, Reading, Mass., 1994.

Fields Institute Communications
Volume **27**, 2000

Recent Results on the Ginzburg-Landau $\nabla\phi$ Interface Model

T. Funaki
Graduate School of Mathematical Sciences
University of Tokyo
3-8-1 Komaba Meguro-ku
Tokyo 153-8914, JAPAN
funaki@ms.u-tokyo.ac.jp

Abstract. This paper reviews recent results on the Ginzburg-Landau $\nabla\phi$ interface model including law of large numbers (LLN), central limit theorem (CLT) and large deviations (LD).

1 Introduction and model

The structure of phase transitions seems still not well understood from the microscopic dynamical point of view. At low temperatures two distinct pure phases may coexist and be separated by a fairly sharp interface. Such a situation will persist under the time evolution and the interface will relax slowly. The goal is to investigate the motion of interface on a properly chosen coarse space-time scale. The prime example is the ferromagnetic Ising model and the corresponding time evolution is a reversible spin-flip dynamics, the so-called Glauber dynamics. Spin at each site randomly flips and changes its sign under the dynamics. At sufficiently low temperatures, the interactions between spins on two neighboring sites become strong enough to incline them to have the common signs with high probability and most changes occur near the interface. The shape of interface is however rather complicated; for instance, it has overhangs. Effective interface model is introduced by avoiding such complications and directly modeling the interface degree of freedom, see [12]. The Ginzburg-Landau $\nabla\phi$ interface model, one of effective interface models, is recently studied by several authors and the aim of this paper is to try to give an overview of the results.

1.1 Summary of recent results. Let us summarize recent results briefly. The model will be explained in the next subsection. The interaction potential V is always supposed strictly convex.

(1) LLN (Section 3): Macroscopic quantity is obtained under suitable scaling limit from randomly fluctuating microscopic objects, i.e. height variables of interfaces in our model. The limit is deterministic.

2000 *Mathematics Subject Classification.* Primary 60K35; Secondary 82B20, 82C20, 35K55.

(1-i) Dynamic LLN: LLN is shown under the time evolution. This procedure is called the hydrodynamic limit and established by Funaki and Spohn [8]. Motion by mean curvature except for some anisotropy is derived in the limit. The diffusion matrix of the equation obtained in the limit is formally given by Hessian of the thermodynamic function called surface tension. Characterization of all Gibbs measures (Section 2, [8]) is required.
(1-ii) Static LLN: Ben Arous and Deuschel [2] studied Gaussian height variables (i.e. variables distributed according to the Gibbs measure with quadratic potential) conditioned to be positive and to have definite total volume. The shape of most probable droplet which minimizes the total surface tension is determined as a consequence of LD result. Generalization to the non-Gaussian case was discussed by Deuschel, Giacomin and Ioffe [5]. See also the paper by Bolthausen and Ioffe [4].

(2) CLT (Section 4): The fluctuations around the deterministic limit are studied.
(2-i) Dynamic CLT in equilibrium: Infinite dimensional Ornstein-Uhlenbeck process is obtained by Giacomin, Olla and Spohn [9]; however, the identification of covariance matrix with Hessian of surface tension still remains open.
(2-ii) Static CLT: Naddaf and Spencer [11] investigated CLT for Gibbs measures. The result is non-trivial since the Gibbs measures have long correlations.

(3) LD (Section 5): LD principles are established for macroscopically scaled height variables.
(3-i) Dynamic LD was discussed by Funaki and Nishikawa [7].
(3-ii) Static LD was studied for Gaussian case by Ben Arous and Deuschel [2] and for general Gibbsian case by Deuschel, Giacomin and Ioffe [5].

1.2 Ginzburg-Landau $\nabla\phi$ interface model. Now let us explain the model. We consider hypersurfaces embedded in $d+1$ dimensional space and assume that they are represented as graphs viewed from a fixed reference hyperplane $\mathbb{R}^d$. The hyperplane $\mathbb{R}^d$ is then discretized and replaced with $\mathbb{Z}^d$. The position of interface is consequently described by the height variables $\phi = \{\phi(x) \in \mathbb{R}, x \in \mathbb{Z}^d\}$; $\phi(x)$ is interpreted as the height of the interface relative to $\phi(x) = 0$.
Hamiltonian: The formal *energy* $H(\phi)$ of the interface ϕ should be invariant under a uniform translation $\phi(x) \to \phi(x) + a$ for all $x \in \mathbb{Z}^d$ and $a \in \mathbb{R}$, and this naturally leads us to

$$H(\phi) = \sum_{\langle x,y\rangle} V(\phi(x) - \phi(y)), \tag{1.1}$$

where the sum is taken over all pairs of nearest neighbor sites $\{x, y\}$ in $\mathbb{Z}^d$. Throughout the paper we require the following conditions on the potential V:

$$\begin{aligned} &\text{(i)} \quad V \in C^2(\mathbb{R}), \\ &\text{(ii)} \quad \text{(symmetry) } V(-\eta) = V(\eta), \quad \eta \in \mathbb{R}, \\ &\text{(iii)} \quad \text{(strict convexity) } c_- \le V''(\eta) \le c_+, \quad \eta \in \mathbb{R}, \quad \text{for some } c_-, c_+ > 0. \end{aligned} \tag{1.2}$$

The surface ϕ has low energy $H(\phi)$ if $|\phi(x) - \phi(y)|$ are small.

Remark 1.1 (i) In the quantum field theory, H is called massless Hamiltonian and well studied in '80s. Massive Hamiltonian is given by $H_m(\phi) := H(\phi) + m\sum_x \phi^2(x), m > 0$ and for massive case the corresponding dynamics have forces for $\phi(x)$ to stay bounded.
(ii) In our model height variables $\phi(x)$ themselves are not discretized. SOS model

is a model obtained discretizing also the height variables: $\phi(x) \in \mathbb{Z}_+$ and with $V(\eta) = |\eta|$.
(iii) Related problems are discussed in [10], [1], [3].

Dynamics: The corresponding time evolution of ϕ-field is governed by the stochastic differential equations (SDEs)

$$d\phi_t(x) = -\frac{\partial H}{\partial \phi(x)}(\phi_t)dt + \sqrt{2}dw_t(x), \quad x \in \mathbb{Z}^d,$$

where $w_t = \{w_t(x), x \in \mathbb{Z}^d\}$ is a family of independent 1-dimensional standard Brownian motions. In other words,

$$d\phi_t(x) = -\sum_{y\in\mathbb{Z}^d:|x-y|=1} V'(\phi_t(x) - \phi_t(y))dt + \sqrt{2}dw_t(x), \quad x \in \mathbb{Z}^d. \tag{1.3}$$

The SDEs (1.3) have unique solutions, since the coefficient V' of the drift term is Lipschitz continuous, see [8] for details. Formally saying, ϕ_t is defined in such a way that it is reversible under Gibbs measures $d\mu = Z^{-1}e^{-H(\phi)}d\phi$. Such evolution and the SDEs are called Ginzburg-Landau dynamics or distorted Brownian motion and the Langevin equation associated with $H(\phi)$, respectively. Note that $-V'(\phi_t(x) - \phi_t(y))$ in the drift term has an effect to make the interface ϕ flat, since $\phi_t(x) - \phi_t(y) > 0 (< 0)$ implies that the drift is negative (positive) so that $\phi_t(x)$ decreases (increases). The term $\sqrt{2}w_t(x)$ gives a random fluctuation. The corresponding Dirichlet form is

$$\mathcal{D}(F,G) \equiv -\langle FLG\rangle_\mu = \sum_x \langle \partial F(x,\phi)\partial G(x,\phi)\rangle_\mu, \tag{1.4}$$

for $F = F(\phi), G = G(\phi)$, where $\langle\cdot\rangle_\mu$ denotes the expectation under μ, L is the generator of the process ϕ_t and $\partial F(x,\phi) := \partial F/\partial\phi(x)$.

2 Gibbs measures

In this section we address the question to characterize all equilibrium (stationary) measures for the dynamics (1.3). This becomes necessary for establishing the hydrodynamic limit.

2.1 $\nabla\phi$-field. Since the dynamics is invariant under the uniform translation $\phi(x) \to \phi(x) + a$, it turns out to be better to consider the dynamics for the height differences (tilts or gradients):

$$\eta_t(b) \equiv \nabla\phi_t(b) := \phi_t(x) - \phi_t(y),$$

where $b = \{x, y\} \in (\mathbb{Z}^d)^*$ are directed bonds in $\mathbb{Z}^d$. Since Gibbs measures for η-field defined later are reversible and therefore stationary for the stochastic processes $\eta_t = \{\eta_t(b), b \in (\mathbb{Z}^d)^*\}$ (see Theorem 2.2), the study of stationary measures for η_t yields information for Gibbs measures. Our result can roughly be stated as follows: Under the condition (1.2) on the potential V, for each $u \in \mathbb{R}^d$ there exists a unique shift invariant, ergodic under shifts and stationary probability measure μ_u for η_t with mean u (average tilt). Especially, there exists a unique Gibbs measure which is shift invariant, ergodic under shifts and has mean u.

2.2 Stationary measures. Let us formulate more precisely. The state space of the process η_t is denoted by

$$\mathcal{X} := \{\eta \in \mathbb{R}^{(\mathbb{Z}^d)^*} \,|\, \eta \text{ satisfies loop condition}\}.$$

Here the loop condition means that $\eta(b) = -\eta(-b)$ for all $b \in (\mathbb{Z}^d)^*$, $-b$ denotes the reversed bond of b, and $\sum_{b\in\mathfrak{C}} \eta(b) = 0$ for all closed loops $\mathfrak{C}$ consisting of bonds in $\mathbb{Z}^d$. We shall consider the following classes of probability measures on $\mathcal{X}$:

$$\begin{aligned}
&\mathcal{P}(\mathcal{X}) = \{\text{probability measures on } \mathcal{X}\},\\
&\mathcal{P}_2(\mathcal{X}) = \{\mu \in \mathcal{P}(\mathcal{X}) |\; E^\mu[\eta^2(b)] < \infty, {}^\forall b \in (\mathbb{Z}^d)^*\},\\
&\mathcal{S} = \{\mu \in \mathcal{P}_2(\mathcal{X}) |\; \text{shift invariant and stationary for } \eta_t\},\\
&\text{ext } \mathcal{S} = \{\mu \in \mathcal{S} |\; \text{ergodic under shifts}\},\\
&(\text{ext } \mathcal{S})_u = \{\mu \in \text{ext } \mathcal{S} |\; \mu \text{ has mean } u\},\; u = (u_\alpha)_{\alpha=1}^d \in \mathbb{R}^d.
\end{aligned}$$

The last condition means $E^\mu[\eta(e_\alpha)] = u_\alpha$ for every $1 \le \alpha \le d$, where $e_\alpha \in \mathbb{Z}^d$ denotes the α-th unit vector given by $(e_\alpha)_\beta = \delta_{\alpha\beta}$. The bond $\{e_\alpha, 0\}$ is also denoted by e_α. The measures $\mu \in \mathcal{P}_2(\mathcal{X})$ are called tempered.

Theorem 2.1 *([8]) (uniqueness)* $\sharp\{(\text{ext}\,\mathcal{S})_u\} \le 1$ *for every* $u \in \mathbb{R}^d$.

The proof of this theorem is given based on a coupling argument; namely, assuming that there exist two different measures $\mu, \bar\mu \in (\text{ext } \mathcal{S})_u$, we construct two solutions ϕ_t and $\bar\phi_t$ of the SDEs (1.3) with common Brownian motions w_t in such a way that the gradients $\nabla\phi_0$ and $\nabla\bar\phi_0$ of their initial data are distributed under μ and $\bar\mu$, respectively. Then, computing the time derivative of $\sum_{x\in\Lambda} E[(\phi_t(x) - \bar\phi_t(x))^2]$ for each finite $\Lambda \subset \mathbb{Z}^d$, one can finally conclude $\mu = \bar\mu$ by letting $\Lambda \nearrow \mathbb{Z}^d$ and noting the ergodicity of μ and $\bar\mu$. The strict convexity of V plays an essential role.

2.3 Gibbs measures. For every $\xi \in \mathcal{X}$ and finite $\Lambda \subset \mathbb{Z}^d$ the space of all possible configurations of height differences on $\overline{\Lambda^*} := \{b = \{x, y\} \in (\mathbb{Z}^d)^* | x$ or $y \in \Lambda\}$ for given boundary condition ξ is defined as

$$\mathcal{X}_{\overline{\Lambda^*},\xi} := \{\eta = (\eta(b))_{b\in\overline{\Lambda^*}} |\, \eta \vee \xi \in \mathcal{X}\},$$

where $\eta \vee \xi \in \mathcal{X}$ is determined by $(\eta\vee\xi)(b) = \eta(b)$ for $b \in \overline{\Lambda^*}$ and $= \xi(b)$ for $b \notin \overline{\Lambda^*}$. The finite volume Gibbs measure in Λ with boundary condition ξ is defined by

$$\mu_{\Lambda,\xi}(d\eta) := Z_{\Lambda,\xi}^{-1} \exp\left\{ -\frac{1}{2} \sum_{b\in\overline{\Lambda^*}} V(\eta(b)) \right\} d\eta_{\Lambda,\xi} \in \mathcal{P}(\mathcal{X}_{\overline{\Lambda^*},\xi}),$$

where $d\eta_{\Lambda,\xi}$ denotes a uniform measure on the affine space $\mathcal{X}_{\overline{\Lambda^*},\xi}$ and $Z_{\Lambda,\xi}$ is the normalization; note that the factor $1/2$ appears here since we now count each (undirected) bonds twice as b and $-b$. $\mu \in \mathcal{P}(\mathcal{X})$ is called a Gibbs measure for the height differences if it satisfies the DLR equations

$$\mu(\cdot|\mathcal{F}_{(\mathbb{Z}^d)^*\setminus\overline{\Lambda^*}})(\xi) = \mu_{\Lambda,\xi}(\cdot), \quad \mu\text{-a.e. } \xi,$$

for every finite $\Lambda \subset \mathbb{Z}^d$, where $\mathcal{F}_{(\mathbb{Z}^d)^*\setminus\overline{\Lambda^*}}$ is the σ-algebra generated by $\{\eta(b) | b \in (\mathbb{Z}^d)^* \setminus \overline{\Lambda^*}\}$.

We shall denote

$$\mathcal{G} = \{\mu \in \mathcal{P}_2(\mathcal{X}) |\, \text{shift invariant Gibbs measures}\},$$

and the classes ext $\mathcal{G}$ and $(\text{ext }\mathcal{G})_u$ are similarly defined, so that $\mu \in \text{ext }\mathcal{G}$ is ergodic under shifts and $\mu \in (\text{ext }\mathcal{G})_u$ has mean u. Since $\mu_{\Lambda,\xi}$ is reversible under the evolution governed by finite dimensional SDEs for η-field on $\overline{\Lambda^*}$ with boundary condition ξ, one can easily show the following theorem by letting $\Lambda \nearrow \mathbb{Z}^d$.

Theorem 2.2 *([8]) Every $\mu \in \mathcal{G}$ is reversible under the dynamics η_t defined by the SDEs (1.3). In particular, we have $\mathcal{G} \subset \mathcal{S}$.*

The next theorem is shown by taking thermodynamic limit (with periodic boundary conditions) again. The tightness of the sequence of finite volume Gibbs measures is a consequence of Brascamp-Lieb inequality which is applicable only to strict convex potentials.

Theorem 2.3 *([8]) (existence) For every $u \in \mathbb{R}^d$, there exists $\mu =: \mu_u \in (\text{ext}\,\mathcal{G})_u$. Furthermore it satisfies $E^{\mu_u}[e^{\beta(\eta(b)-u_b)^2}] < \infty$ for some $\beta > 0$, where $u_b = u_\alpha$ if b is α-directed.*

Theorems 2.1 and 2.2 imply the uniqueness of ergodic Gibbs measures for each mean tilt u and therefore combining with Theorem 2.3 we have the following corollary.

Corollary 2.4 *([8]) (characterization of all tempered Gibbs measures) $\mathcal{G}$ is the convex hull of $\{\mu_u | \, u \in \mathbb{R}^d\}$.*

The well-known Dobrushin's uniqueness argument does not work here, since if it works the correlation functions decay exponentially fast and this may contradict with the next remark (ii).

Remark 2.5 (i) We have only considered the Gibbs measures for η-field. The Gibbs measure for ϕ-field is unnormalizable (i.e. finite volume Gibbs measures for ϕ-field do not converge as $\Lambda \nearrow \mathbb{Z}^d$) if $d \le 2$ and normalizable if $d \ge 3$.
(ii) The correlation functions of Gibbs measures decay slowly. For example, if $V(\eta) = \eta^2/2$, then $H(\phi) = -(\Delta\phi, \phi)_{\ell^2(\mathbb{Z}^d)}/2$ where Δ denotes the discrete Laplacian and therefore the Gibbs measure for ϕ-field is Gaussian with covariance $\langle \phi(x); \phi(y)\rangle_\mu := \langle (\phi(x) - \langle \phi(x)\rangle_\mu)(\phi(y) - \langle \phi(y)\rangle_\mu)\rangle_\mu = G(x,y), x, y \in \mathbb{Z}^d$, the Green function of simple random walk on $\mathbb{Z}^d$. Note that $G(x,y)$ is finite iff $d \ge 3$ and behaves $G(x,y) \sim k/|x-y|^{d-2}$ as $|x-y| \to \infty$. In this sense, the random field has long dependence and this makes the proof of CLT or LD difficult. Naddaf and Spencer [11] proved for general convex potential V

$$|\langle \phi(x); \phi(y)\rangle_\mu| \le \frac{C}{|x-y|^{d-2}}$$

if $d \ge 3$. One can say that the loop condition for η-field gives the long dependence.

3 Hydrodynamic limit

Let us consider the SDEs (1.3) on a big but finite periodic cubic lattice $\Gamma_N := (\mathbb{Z}/N\mathbb{Z})^d = \{1, 2, \ldots, N\}^d$ (we identify $N+1$ with 1):

$$d\phi_t(x) = -\sum_{y \in \Gamma_N : |x-y|=1} V'(\phi_t(x) - \phi_t(y))dt + \sqrt{2}dw_t(x), \quad x \in \Gamma_N. \tag{3.1}$$

We work on a periodic lattice to avoid thinking about the boundary condition.

3.1 Diffusive scaling. Introduce a diffusive scaling to study the macroscopic behavior of height variables. Macroscopic height variables for the interface is defined as a step function on the torus $\mathbb{T}^d := (\mathbb{R}/\mathbb{Z})^d = [0,1)^d$ as follows:

$$\phi_t^{(N)}(\theta) := \frac{1}{N}\phi_{N^2t}(x), \quad \theta \in B(x/N, 1/N), \tag{3.2}$$

where $B(\theta, a) = \prod_{\alpha=1}^d [\theta_\alpha - a/2, \theta_\alpha + a/2)$ denotes a box in $\mathbb{T}^d$ with center $\theta = (\theta_\alpha)_{\alpha=1}^d$ and side length $a > 0$. Note that both x- and ϕ-axes are rescaled by a factor $1/N$. This is because the ϕ-field represents a hypersurface embedded in $d+1$ dimensional space.

Theorem 3.1 ([8]) *Assume that initial random configuration ϕ_0 of the SDEs (3.1) converges to some non-random $h_0 \in L^2(\mathbb{T}^d)$ in the sense that*

$$\lim_{N\to\infty} E[||\phi_0^{(N)} - h_0||^2] = 0, \tag{3.3}$$

where $||\cdot||$ denotes the usual L^2-norm of the space $L^2(\mathbb{T}^d)$. Then, for every $t > 0$,

$$\lim_{N\to\infty} E[||\phi_t^{(N)} - h_t||^2] = 0$$

holds and $h_t = h(t,\theta)$ is a unique solution of the nonlinear partial differential equation (PDE)

$$\frac{\partial}{\partial t}h(t,\theta) = \operatorname{div}[(\nabla\sigma)(\nabla h(t,\theta))], \quad \theta \in \mathbb{T}^d, \tag{3.4}$$

having initial data h_0.

If $\sigma \in C^2(\mathbb{R}^d)$ which is not yet shown, see Theorem 3.2-(ii) and Open Problem below, the diffusion coefficient of the PDE (3.4) is given by the Hessian $(\partial^2\sigma/\partial u_\alpha \partial u_\beta)_{\alpha,\beta}$ of σ.

3.2 Surface tension. The function $\sigma = \sigma(u)$ appearing in the equation (3.4) is called surface tension which is determined as a thermodynamic function from the Hamiltonian $H(\phi)$ and reflects a statistical property of random interfaces with mean tilt $u \in \mathbb{R}^d$. Let

$$\sigma_\Lambda(u) := -\frac{1}{|\Lambda|}\log \int \exp\{-H_{\Lambda,u}(\phi)\} \prod_{x\in\Lambda\setminus\partial\Lambda} d\phi(x), \qquad u \in \mathbb{R}^d,$$

where $H_{\Lambda,u}(\phi) = \sum_{\langle x,y\rangle\subset\Lambda} V(\phi(x)-\phi(y))$ with $\phi(y) = u\cdot y$ for $y \in \partial\Lambda$; note that the tilt u is insisted through the boundary condition.

Theorem 3.2 ([8]) *(i) The limit $\sigma(u) := \lim_{\ell\to\infty}\sigma_{\Lambda_\ell}(u)$ exists, where $\Lambda_\ell = [-\ell,\ell]^d \cap \mathbb{Z}^d$.*
(ii) The function σ is strictly convex ([5], [9]), symmetric, $\sigma \in C^1(\mathbb{R}^d)$ and $\nabla\sigma$ is Lipschitz continuous.

The function $\sigma(u)$ gives an energy of a (macroscopic) surface with tilt u; in particular, surface with gentle slope has low energy. The assertion (i) of Theorem 3.2 follows from the subadditivity of $\sigma_{\Lambda_\ell}(u)$ in ℓ. The strict convexity of σ is a consequence of uniform strict convexity of σ_{Λ_ℓ} in ℓ. The Lipschitz continuity of $\nabla\sigma$ is shown based on the coupling used for the proof of Theorem 2.1 with the help of (3.6) below.

Theorem 3.3 *([8]) (thermodynamic identities)*

$$E^{\mu_u}[\eta(e_\alpha)] = u_\alpha, \tag{3.5}$$

$$E^{\mu_u}[V'(\eta(e_\alpha))] = \frac{\partial\sigma}{\partial u_\alpha}(u), \tag{3.6}$$

$$E^{\mu_u}\left[\sum_{\alpha=1}^{d}\eta(e_\alpha)V'(\eta(e_\alpha))\right] = u\cdot\nabla\sigma(u)+1. \tag{3.7}$$

The identity (3.5) is just by definition, while (3.6) and (3.7) are shown by the integration by parts for similar integrals appearing in σ_Λ.

Remark 3.4 (i) Set $\mathbb{A}(h) := \mathrm{div}[\nabla\sigma(\nabla h)]$ the right hand side of the nonlinear PDE (3.4). Then the convexity of σ implies the monotonicity of $\mathbb{A} : H^1(\mathbb{T}^d) \to H^{-1}(\mathbb{T}^d)$ and $u\cdot\nabla\sigma(u) \ge c_-|u|^2 - 1$ which follows from (3.7) shows its coercivity. The existence and uniqueness of (weak) solutions to the PDE (3.4) follows from these properties of the operator $\mathbb{A}$.
(ii) (Physical argument on $\sigma(u)$) See [6] for roughening transition: $\sigma \notin C^1$ and has cusp at $u = 0$. Phase transition is expected for general non-convex potentials V.
(iii) In the Gaussian case (i.e. $V(\eta) = \eta^2/2$), computing $\sigma_{\Lambda_\ell}(u)$ explicitly, we have $\sigma(u) = |u|^2/2$ and the limit equation (3.4) is linear heat equation. In fact, this can be directly seen, since the drift term of the SDEs (3.1) is $\Delta\phi_t(x)$ for such potential V; recall that Δ denotes the discrete Laplacian.

3.3 Physical meaning of the PDE (3.4). Consider the total surface tension (surface free energy) of the macroscopic surface $h = \{h(\theta), \theta \in \mathbb{T}^d\}$:

$$\mathbb{F}(h) := \int_{\mathbb{T}^d}\sigma(\nabla h(\theta))\,d\theta. \tag{3.8}$$

Then, its Fréchet derivative is given by

$$\frac{\delta\mathbb{F}}{\delta h(\theta)}(h) = -\,\mathrm{div}[(\nabla\sigma)(\nabla h(\theta))],$$

Therefore the hydrodynamic equation (3.4) is a gradient flow for $\mathbb{F}(h)$:

$$\frac{\partial h}{\partial t} = -\frac{\delta\mathbb{F}}{\delta h}, \tag{3.9}$$

namely the surface moves relaxing its total surface energy. For isotropic motion by mean curvature one would have $\sigma(u) = \sqrt{1+|u|^2}$, note that $\mathbb{F}(h)$ is the surface area of h and $\delta\mathbb{F}/\delta h$ is the mean curvature in such case. In our case this is likely to hold for small $|u|$, however $\sigma(u) \simeq |u|^2$ for large $|u|$, which reflects the constraints due to the underlying microscopic lattice structure.

3.4 Formal derivation of the PDE (3.4). For every test function $\varphi = \varphi(\theta) \in C^\infty(\mathbb{T}^d)$,

$$\begin{aligned}\langle\phi_t^{(N)},\varphi\rangle :=& \int_{\mathbb{T}^d}\phi_t^{(N)}(\theta)\varphi(\theta)\,d\theta\\ =&N^{-d}\sum_{x\in\Gamma_N}\phi_t^{(N)}(x/N)\varphi(x/N)+o(1)\end{aligned}$$

as $N \to \infty$. Then, applying Itô's formula and recalling the symmetry of V, we have

$$d\langle \phi_t^{(N)}, \varphi \rangle = -N^{-d} \sum_{\alpha=1}^{d} \sum_{x \in \Gamma_N} V'(\nabla \phi_{N^2 t}(x+e_\alpha)) \frac{\partial \varphi}{\partial \theta_\alpha}(x/N)\, dt + o(1).$$

The last error $o(1)$ contains martingale term. The left hand side would converge to $d\langle h_t, \varphi \rangle$. On the other hand, by means of the local ergodicity,

$$V'(\nabla \phi_{N^2 t}(x+e_\alpha))$$

in the right hand side could be replaced with its ensemble average

$$E^{\mu_{\nabla h_t(x/N)}}[V'(\nabla \phi(e_\alpha))]$$

which coincides with $\partial \sigma / \partial u_\alpha (\nabla h_t(x/N))$ from (3.6). Therefore, one would obtain in the limit

$$\frac{d}{dt} \langle h_t, \varphi \rangle = -\langle \nabla \sigma(\nabla h_t), \nabla \varphi \rangle,$$

for every $\varphi \in C^\infty(\mathbb{T}^d)$, which is a weak form of the nonlinear PDE (3.4). The actual proof given in [8] is slightly different. It is based on the H^{-1}-method, by which one can avoid the so-called two blocks' estimate. Corollary 2.4 is required to complete the proof.

3.5 Static LLN. Ben Arous and Deuschel [2] studied the Gaussian random field $\{\phi(x), x \in [0,N]^d \cap \mathbb{Z}^d\}$ distributed under the finite volume Gibbs measure for $V(\eta) = \eta^2/2$ satisfying $\phi(x) \equiv 0$ at the boundary. Then, the field is conditioned in such a manner that $\phi(x) \geq 0$ and macroscopic total volume $= v$ (i.e. $N^{-d} \sum_x \phi(x) = v$). They proved for the conditioned field that on several kind of scalings the macroscopic height variables $\phi^{(N)}(\theta), \theta \in D = [0,1]^d$ converge as $N \to \infty$ to $h^*(\theta)$ which minimizes the total surface tension $\int_D |\nabla h|^2 d\theta / 2d$ under the condition $h = 0$ at ∂D, $h \geq 0$ and $\int_D h d\theta = v$. Recent paper by Deuschel, Giacomin and Ioffe [5] generalized the results to the non-Gaussian setting.

4 Fluctuations

4.1 Dynamic CLT. Let $\{\phi_t(x), x \in \mathbb{Z}^d\}$ be the solution of the SDEs (1.3) with initial data ϕ_0 whose gradients $\nabla \phi_0$ are distributed according to μ_u for some $u \in \mathbb{R}^d$. Note that the process $\eta_t \equiv \nabla \phi_t$ is in equilibrium. Consider $\mathcal{S}'(\mathbb{R}^d)$-valued process

$$\xi_{t,\alpha}^N(d\theta) := N^{-d/2} \sum_{x \in \mathbb{Z}^d} (\eta_{N^2 t}(x+e_\alpha) - u_\alpha)\, \delta_{x/N}(d\theta), \qquad \theta \in \mathbb{R}^d,$$

for each $1 \leq \alpha \leq d$.

Theorem 4.1 *([9]) The process $\xi_{t,\alpha}^N$ weakly converges in the space $C([0,\infty), \mathcal{S}'(\mathbb{R}^d))$ to a solution $\xi_{t,\alpha}$ of an infinite dimensional SDE*

$$d\xi_{t,\alpha}(\theta) = -A\xi_{t,\alpha}(\theta)\, dt + \sqrt{2} \frac{\partial}{\partial \theta_\alpha} dw_t(\theta),$$

where

$$A = -\sum_{\alpha,\beta=1}^{d} q_{\alpha,\beta} \frac{\partial^2}{\partial \theta_\alpha \partial \theta_\beta},$$

with a positive definite $d \times d$ matrix $(q_{\alpha,\beta}) \equiv (q_{\alpha,\beta}(u))$ characterized by a certain variational formula and $\dot{w}_t(\theta)$ is a space-time white noise.

The proof relies on the Helffer-Sjöstrand type representation (see Subsection 4.2) for correlation functions of the process η_t in terms of random walk in moving random environment. Such representation is also used in [11] for static correlation functions. Then the problem is reduced to establishing a homogenization for this random walk.

Remark 4.2 Theorem 4.1 covers the static CLT shown by Naddaf and Spencer [11].

Open Problem ([9]): Does the strict convexity of V imply $\sigma \in C^2(\mathbb{R}^d)$ and $q(u) =$ Hessian of $\sigma(u)$? Here $q(u) = (q_{\alpha,\beta}(u))$ is the covariance matrix appearing in CLT, while $\sigma(u)$ is the surface tension obtained in LLN. Such identity usually holds, for instance if the Gibbs measures have exponentially fast mixing property.

4.2 H-S representation. The Helffer-Sjöstrand representation gives the following identity for the covariance of $F = F(\phi)$ and $G = G(\phi)$ under the Gibbs measure μ:

$$\langle F; G\rangle_\mu = \sum_x \int_0^\infty \langle \partial F(x,\phi_0)\partial G(X_t,\phi_t)\rangle_\mu \, dt, \tag{4.1}$$

where ϕ_0 is μ-distributed, X_t is the random walk on $\mathbb{Z}^d$ starting at x with generator Q_{ϕ_t} defined by

$$Q_\phi j(x) = \sum_{y:|x-y|=1} V''(\phi(x)-\phi(y))\{j(y)-j(x)\}, \qquad j:\mathbb{Z}^d \to \mathbb{R}.$$

Indeed, assume $\langle G\rangle_\mu = 0$ and let H be the solution of the Poisson equation $-LH = G$. Then, from (1.4)

$$\langle F; G\rangle_\mu = -\langle FLH\rangle_\mu = \sum_x \langle \partial F(x,\phi)\partial H(x,\phi)\rangle_\mu .$$

However, a simple computation shows

$$\partial(LH)(x,\phi) = \{(L+Q_\phi)\partial H\}(x,\phi)$$

and therefore

$$\partial H(x,\phi) = (L+Q_\phi)^{-1}\partial(LH)(x,\phi) = \int_0^\infty E_{(x,\phi)}\left[\partial G(X_t,\phi_t)\right] dt.$$

This implies the identity (4.1). The above argument is rather formal and, in particular, one should replace the measure μ with finite volume Gibbs measure [5] or with Gibbs measure for η-field [9].

5 Large deviations

5.1 Dynamic LD. The hydrodynamic limit is an LLN result for the macroscopic height variables $\phi_t^{(N)}$ defined in (3.2). In this section we mention the corresponding LD principle. We always assume the condition (1.2) on the potential V. The result is the following: Assume that the initial data $\{\phi_0(x) \equiv \phi_{0,N}(x), x \in \Gamma_N\}$ of the SDEs (3.1) are deterministic and satisfy

$$\sup_N \left\{ |\phi_{0,N}(0)| + N^{-d} \sum_{b\in\Gamma_N^*} (\nabla\phi_{0,N}(b))^2 \right\} < \infty. \tag{5.1}$$

We also assume the condition (3.3) without taking expectations for the corresponding $\phi_0^{(N)}$ and some $h_0 \in L^2(\mathbb{T}^d)$.

Theorem 5.1 *([7]) LD principle holds for $\{\phi_t^{(N)}, t \in [0,T]\}$ with rate functional $I_T(h)$, namely, for every closed subset $\mathcal{C}$ and open subset $\mathcal{O}$ of $C_w([0,T], L^2(\mathbb{T}^d))$, we have*

$$\limsup_{N\to\infty} N^{-d} \log P(\phi^{(N)}_\cdot \in \mathcal{C}) \le - \inf_{h\in\mathcal{C}} I_T(h), \tag{5.2}$$

$$\liminf_{N\to\infty} N^{-d} \log P(\phi^{(N)}_\cdot \in \mathcal{O}) \ge - \inf_{h\in\mathcal{O}} I_T(h), \tag{5.3}$$

where $C_w([0,T], L^2(\mathbb{T}^d))$ stands for the class of all weakly continuous functions $h : [0,T] \to L^2(\mathbb{T}^d)$.

5.2 Dynamic rate functional. For each $h = h(t,\theta)$ which is differentiable,

$$I_T(h) := \frac{1}{4}\int_0^T dt \int_{\mathbb{T}^d} \left\{ \frac{\partial h}{\partial t}(t,\theta) - \operatorname{div}[(\nabla\sigma)(\nabla h(t,\theta))] \right\}^2 d\theta, \tag{5.4}$$

if $h(0) = h_0$ and $I_T(h) = +\infty$ if $h(0) \ne h_0$, where σ is the surface tension defined in Theorem 3.2. More precisely saying, for $h \in C_w([0,T], L^2(\mathbb{T}^d))$ satisfying $h(t) \in H^1(\mathbb{T}^d)$ for every $t \in [0,T]$

$$I_T(h) = \sup_{J=J(t,\theta)\in C^1([0,T]\times\mathbb{T}^d)} I_T(h;J),$$

where

$$\begin{aligned} I_T(h;J) = &\int_{\mathbb{T}^d} J(T,\theta)h(T,\theta)\,d\theta - \int_{\mathbb{T}^d} J(0,\theta)h_0(\theta)\,d\theta - \int_0^T dt \int_{\mathbb{T}^d} \frac{\partial J}{\partial t}(t,\theta)h(t,\theta)\,d\theta \\ &+ \int_0^T dt \int_{\mathbb{T}^d} \nabla J(t,\theta)\cdot\nabla\sigma(\nabla h(t,\theta))\,d\theta - \int_0^T dt \int_{\mathbb{T}^d} J^2(t,\theta)\,d\theta, \end{aligned}$$

and $H^1(\mathbb{T}^d)$ denotes the Sobolev space on $\mathbb{T}^d$.

The upper bound is shown based on the exponential Chebyshev's inequality, while for the lower bound hydrodynamic limit for weakly perturbed system is established and then Girsanov's formula is applied. Essential role is played by the superexponential estimate, namely the probability of replacing sample mean over box of side length $N\epsilon$ with ensemble mean is superexponentially small as $N \to \infty$ and then $\epsilon \downarrow 0$. H^{-1}-method is effectively used to prove such estimate.

5.3 Static LD. Let μ_N be the finite volume Gibbs measure for scaled ϕ-field on $D \cap \frac{1}{N}\mathbb{Z}^d$ with 0-boundary condition, where D is a bounded domain in $\mathbb{R}^d$ with Lipschitz boundary. Deuschel, Giacomin and Ioffe [5] proved LD principle for μ_N with rate functional

$$\mathbb{F}_D(\bar{h}) := \int_D \sigma(\nabla\bar{h}(\theta))\,d\theta, \tag{5.5}$$

the total surface tension of the surface $\bar{h} = \{\bar{h}(\theta), \theta \in D\}$. The Gaussian case was discussed by Ben Arous and Deuschel [2], Theorem 2.5. The measure μ_N has long dependence (see Remark 2.5-(ii)) and therefore the proof is non-trivial. The main tool for the proof is Helffer-Sjöstrand representation.

Since the distribution of $\nabla\phi_T^{(N)}$ weakly converges as $T \to \infty$ to the finite volume Gibbs measure on Γ_N for $\nabla\phi$-field, the natural question is whether one can recover the static LD from the dynamic LD. An affirmative answer is not known at present, however one can at least recover the static rate functional from the dynamic one as $T \to \infty$. In fact, denoting the distribution of $\phi_T^{(N)}$ by μ_T^N, the contraction principle implies the LD principle for $\{\mu_T^N\}_N$ with rate functional

$$S_T(\bar{h}) := \inf_{h=h(t,\theta) \text{ s.t. } h(T,\theta)=\bar{h}(\theta)} I_T(h), \qquad \bar{h} = \bar{h}(\theta) \in H^1(\mathbb{T}^d).$$

The relationship between $S_T(\bar{h})$ and the total surface tension $\mathbb{F}(\bar{h})$ on the torus (see (3.8)) is stated in the following proposition:

Proposition 5.2

$$\lim_{T\to\infty} S_T(\bar{h}) = \mathbb{F}(\bar{h}).$$

The limit in the left hand side is called a quasi-potential and the corresponding classical flow is a gradient flow for the potential $\mathbb{F}$ as pointed out in Section 3. In such case the quasi-potential coincides with the potential $\mathbb{F}$ itself. The infimum is attained by the reversed trajectory of the gradient flow.

References

[1] D. Abraham, P. Collet, J. De Coninck and F. Dunlop, *Langevin dynamics of an interface near a wall*, J. Statis. Phys., **61** (1990), pp. 509-532.

[2] G. Ben Arous and J.-D. Deuschel, *The construction of the d+1-dimensional Gaussian droplet*, Commun. Math. Phys., **179** (1996), pp. 467-488.

[3] E. Bolthausen, J.-D. Deuschel and O. Zeitouni, *Entropic repulsion of the lattice free field*, Commun. Math. Phys., **170** (1995), pp. 417-443.

[4] E. Bolthausen and D. Ioffe, *Harmonic crystal on the wall: a microscopic approach*, Commun. Math. Phys., **187** (1997), pp. 523-566.

[5] J.-D. Deuschel, G. Giacomin and D. Ioffe, *Large deviations and concentration properties for $\nabla\varphi$ interface models*, preprint, 1999.

[6] J. Fröhlich and T. Spencer, *The Kosterlitz-Thouless transition in two-dimensional abelian spin systems and the Coulomb gas*, Commun. Math. Phys., **81** (1981), pp. 527-602.

[7] T. Funaki and T. Nishikawa, *Large deviations for the Ginzburg-Landau $\nabla\phi$ interface model*, preprint, 1999.

[8] T. Funaki and H. Spohn, *Motion by mean curvature from the Ginzburg-Landau $\nabla\phi$ interface model*, Commun. Math. Phys., **185** (1997), pp. 1-36.

[9] G. Giacomin, S. Olla and H. Spohn, *Equilibrium fluctuations for $\nabla\varphi$ interface model*, preprint, 1999.

[10] J. L. Lebowitz and C. Maes, *The effect of an external field on an interface, entropic repulsion*, J. Statis. Phys., **46** (1987), pp. 39-49.

[11] A. Naddaf and T. Spencer, *On homogenization and scaling limit of some gradient perturbations of a massless free field*, Commun. Math. Phys., **183** (1997), pp. 55-84.

[12] H. Spohn, *Interface motion in models with stochastic dynamics*, J. Statis. Phys., **71** (1993), pp. 1081-1132.

Fields Institute Communications
Volume **27**, 2000

Large Scale Behavior of a System of Interacting Diffusions

Ilie Grigorescu
Department of Mathematics
University of Utah
155 South 1400 East, JWB 221
Salt Lake City, Utah 84112-0900
igrigore@math.utah.edu

Abstract. We present the explicit derivation of the asymptotic law of the tagged particle process for a system of interacting Brownian motions in the presence of a diffusive scaling in non-equilibrium. The interaction is local and interpolates between the totally independent case (non-interacting) and the totally reflecting case. It can be viewed as the limiting local version of an interaction through a pair potential as its support shrinks to zero. We also prove the independence of two tagged particles in the limit and analyze the case when the initial profile is singular.

1 Introduction

Assume we have decided to look at a given interacting particles system. There are various scaling options for the same dynamics, but we are interested in the nontrivial ones, i.e. those when certain quantities such as spatial averages are conserved. A system of $n \in Z_+$ independent Brownian motions $(x_1(t), x_2(t), \ldots, x_n(t))_{t\geq 0}$ evolving on a circle of size $N > 0$ can be scaled down to $\frac{1}{N}x_i(t)$, for $i = 1, 2, \ldots, n$ to produce a n-dimensional process on the unit circle. The number of particles can be chosen such that $n = \bar{\rho}N$, for $\bar{\rho} > 0$. The average density on the circle stays $\bar{\rho}$ for any $N > 0$. For a meaningful diffusive evolution we shall speed up time by a factor of N^2. The resulting dynamics is a set of $n = \bar{\rho}N$ independent Brownian motions $(x_1^N(t), x_2^N(t), \ldots, x_n^N(t))_{t\geq 0}$ on the n-dimensional unit torus Γ^n denoted by P^N.

The scaled process represents a *microscopic* realization of the interaction. By contrast, we shall look at *macroscopic* quantities. The average density $\bar{\rho}$ is the only preserved macroscopic quantity in our system. At time $t = 0$ we assume the existence of a measure $\mu(0, dx)$ on the unit circle with total mass $\mu(0, \Gamma) = \bar{\rho}$ such

---

2000 *Mathematics Subject Classification.* Primary 60K35; Secondary 82C22, 82C05.
Research supported by McMaster University and the Fields Institute of Research in Mathematical Sciences.

that

$$\lim_{N\to\infty} \frac{1}{N}\sum_{i=1}^{n} \delta_{x_i(0)} = \mu(0,dx) \tag{1.1}$$

in weak sense. This macroscopic quantity is the initial profile of the system. A natural quantity to look at is the *hydrodynamic* limit:

$$\lim_{N\to\infty} \frac{1}{N}\sum_{i=1}^{n} \delta_{x_i(t)} = \mu(t,dx) \quad . \tag{1.2}$$

For independent Brownian motions this amounts to the classic law of large numbers. The measure-valued process $\{\mu(t,dx)\}_{t\geq 0}$ will be concentrated on the unique deterministic solution of the heat equation $\rho_t = \frac{1}{2}\rho_{xx}$ in the distribution sense with initial condition $\lim_{t\to 0}\rho(t,x)dx = \mu(0,dx)$. Large deviations from the macroscopic profile are studied for this model in [7]. We are also interested in the existence of a limiting measure Q^μ on the path space $\Omega = C([0,\infty),\Gamma)$ with initial distribution $Q^\mu(\omega(0)\in dx) = \mu(0,dx)$ such that

$$\lim_{N\to\infty} \frac{1}{N}\sum_{i=1}^{n} \delta_{x_i(\cdot)} = Q^\mu \quad . \tag{1.3}$$

Again, for the non-interacting case, the limit Q^μ is identical to a Wiener measure on the unit circle starting at $\mu(0,dx)$.

A third quantity we may look at is the fluctuations from the hydrodynamic limit. Let $\varphi(x)$ and $\psi(x)$ be smooth test functions on Γ. The random field $\{\zeta_N(t)\}_{t\geq 0}$ with values in $H^{-2}(\Gamma)$, defined in equilibrium by

$$\langle \varphi, \zeta_N(t)\rangle = \frac{1}{\sqrt{N}}\sum_{i=1}^{n}(\varphi(x_i(t)) - \int_\Gamma \varphi(x)dx) \tag{1.4}$$

converges to a generalized Ornstein-Uhlenbeck process $\{\xi(t)\}_{t\geq 0}$ satisfying

$$d\zeta(t) = \frac{1}{2}\partial_{xx}\zeta(t)dt + \sqrt{\bar\rho}(d\partial_x W(t)) \quad . \tag{1.5}$$

Here $\partial_x W(t)$ is the Gaussian random field with covariance

$$E[\langle\varphi,\partial_x W(t)\rangle\langle\psi,\partial_x W(s)\rangle] = \min\{s,t\}\int_\Gamma \varphi'(x)\psi'(x)dx \ .$$

The averaging taking place in the law of large numbers (1.3) may wipe out a lot of features of the interaction between particles. However, in the independent case we have in mind, the limit Q^μ is essentially the same as the law of each particle considered by itself, for a fixed label, (say equal to one) $\{x_1(t)\}_{t\geq 0}$. This is the *tagged particle* process or the so-called tracer particle. The individual particles have trajectories supported on the set of continuous paths on the unit circle Γ. Every such path can pe lifted in a canonical way to the real line. Theorems 2.3, 2.4 are true for the lifted process. However we shall not differentiate between the two in the context of this paper.

The outlook of these considerations changes radically as soon as we allow some interaction between the particles in the system. Suppose $V(x)$ is a smooth, positive, even potential with compact support. Let's consider the n-dimensional interacting Brownian motions process P_V^N with generator

$$\mathcal{L}f(x) = \frac{1}{2}\Delta f(x) - \sum_i (\sum_{j\neq i} V'(x_i - x_j))\partial_i f(x) \text{ with } \mathcal{D}(\mathcal{L}) = C^2(\Gamma^n, R) \quad . \tag{1.6}$$

The presence of the Lapalcian induces a diffusive scaling $t_{micro} = N^2 t_{macro}$ and $x_{micro} = N x_{macro}$. This amounts to the scaling of the potential $V_N(x) = V(Nx)$. The equilibrium measure is given by the density

$$\frac{1}{Z_N} \exp\left[-\frac{1}{2}\sum_{i,j} V(N(x_i - x_j))\right] \quad . \tag{1.7}$$

Under the initial condition (1.1) one can obtain a hydrodynamic limit for the macroscopic density of the system. The bulk diffusion appears in [5]. See [18] for the nonequilibrium case. The limit for the fluctuation field from the macroscopic density in equilibrium is a Gaussian random field of the same nature as (1.5), shown in [13]. (See [1] for the Ginzburg-Landau model in nonequilibrium.)

The tagged particle process has been studied in equilibrium by Guo. The self-diffusion coefficient has a variational formula ([4] and [6]). The self-diffusion coefficient in nonequilibrium has not been done.

Another type of interaction we may be interested in studying is an extreme case of (1.6). At this level we are not interested in the scaling for the interacting particle system. The primary fact is that for a fixed n and any $\epsilon > 0$ we can consider a potential $V_\epsilon(x)$ of the same nature but with support included in the interval $[-\epsilon, \epsilon]$. Assume there exists a positive λ such that

$$\lim_{\epsilon \to 0} \int_\Gamma (\exp(2V_\epsilon(x)) - 1)dx = \frac{1}{\lambda} \quad . \tag{1.8}$$

Then $\{P_{V_\epsilon}^N\}_\epsilon$ is tight and has a unique limit P_λ^N which is described below.

2 The interaction model

Consider a positive integer n and $\lambda \geq 0$. Let Γ^n be the n-dimensional torus. We define $F^{ij} = \{x \in \Gamma^n : x_i = x_j\}$ for any i, j in $\{1, \ldots, n\}$ and $F = \cup_{1\leq i<j\leq n} F_{ij}$. With this notation we introduce the following class of functions.

Definition 2.1 Let $\bar{C}(\Gamma^n, F) = \{f : \Gamma^n \longrightarrow R : f \in C^2(\Gamma^n \setminus F)$ with $f^{ij}(x_0)$ and $D^{ij}f(x_0)$ finite for any $x_0 \subset F$ and any $(i,j)\}$ - the set of smooth functions up to the boundary F, where f^{ij} and $D^{ij}f$ are defined as

$$\begin{aligned} &f^{ij}(x_1, x_2, ..., x_{i-1}, x, x_{i+1}, ..., x_{j-1}, x, x_{j+1}, ..., x_n) \\ &= f(x_1, x_2, ..., x_{i-1}, x+0, x_{i+1}, ..., x_{j-1}, x-0, x_{j+1}, ..., x_n) \end{aligned} \tag{2.1}$$

and

$$\begin{aligned} &D^{ij} f(x_1, x_2, ..., x_{i-1}, x, x_{i+1}, ..., x_{j-1}, x, x_{j+1}, ..., x_n) \\ &= (\partial_i - \partial_j) f(x_1, x_2, ..., x_{i-1}, x+0, x_{i+1}, ..., x_{j-1}, x-0, x_{j+1}, ..., x_n) \quad . \end{aligned} \tag{2.2}$$

We are now in a position to define the generator of the process

$$x^n(t) = (x_1(t), \ldots, x_n(t))$$

on Γ^n. For a real $\lambda \geq 0$ we define the boundary conditions:

$$(BC) \qquad D^{ij}f(x) + \lambda(f^{ji}(x) - f^{ij}(x)) = 0 \quad \forall i,j \in \{1,\ldots,n\} \quad . \tag{2.3}$$

The operator $(\mathcal{L},\mathcal{D}(\mathcal{L}))$ with

$$\mathcal{L}f = \frac{1}{2}\Delta f \quad , \qquad \mathcal{D}(\mathcal{L}) = \{f \in \bar{C}(\Gamma^n, F) : \text{(BC) are satisfied}\} \tag{2.4}$$

is the infinitesimal generator of a process P_λ^n on Γ^n.

This interaction model can be more easily understood by considering a set of canonical local times $A^{ij}(t)$, defined for any oriented pair (i,j), $i \neq j$, $(i,j) \in \{1,\ldots,n\}^2$ which measure the collision time of two particles x_i and x_j when x_i approaches x_j in positive trigonometric order $x_i > x_j$ on the unit circle. The ordering is nonambiguous because we are only interested in the circumstance when the particles are close to each other. In this framework, we shall consider increasing sequences of Poisson events of intensity λ running according to the local times for each oriented pair (i,j). These will be the "jump" times, or, more accurately, the times of switching of labels. Two particles $x_i(t)$ and $x_j(t)$ with distinct labels i and j will perform independent Brownian motions until they meet. Let τ be an exponential time with intensity λ, independent from the process. We define the stopping time $t_{ij} = \inf_{t>0}\{A^{ij}(t) > \tau\}$. The two particles will bounce back from each other according to a reflected Brownian motion. However, once the time t_{ij} is reached, they exchange labels or, in other words, cross each other. This construction repeats itself in the larger setting of the n particle process. The dynamics is that of reflected Brownian motions until the first "jump" time occurs. The particles will continue to perform a n-dimensional reflected Brownian motion until the next exchange of labels. This evolution interpolates between the reflected Brownian motion ($\lambda = 0$) and the independent case ($\lambda = \infty$). The diffusive scaling for the present interaction model amounts to blowing up the parameter λ by a factor of N to $\lambda_N = N\lambda$.

The new scaled process, denoted by P^N, allows us to calculate explicitly the self-diffusion coefficient in *nonequilibrium*. In the equilibrium case the self-diffusion coefficient can be expressed in terms of a variational formula for various interacting particle systems, as in [8], [14], [16] and [17].

Let $\mu(s,dx) = \rho(s,x)dx$ for $s > 0$ be the solution to the heat equation, with initial conditions (1.1). The next theorem is a version of the hydrodynamic limit valid uniformly in time.

Theorem 2.2 *For any smooth $f : \Gamma \longrightarrow R$ and any $t > 0$*

$$\lim_{N\to\infty} E^N \sup_{0\leq s\leq t} \left| \frac{1}{N} \sum_{k=1}^{n=N\bar{\rho}} f(x_k^N(s)) - \int_0^1 f(x)\mu(s,dx) \right|^2 = 0 \quad . \tag{2.5}$$

The hydrodynamic limit of the system is identical to the macroscopic profile of the independent Brownian motions (1.2). The symmetry of the boundary conditions cancels out any trace of the switching of labels occurring microscopically.

The invariant measure of the process is the product uniform distribution on the torus Γ^n. In equilibrium the macroscopic profile will be constant equal to $\bar{\rho}$ and the tagged particle process $\{x_1^N(t)\}_{t\geq 0}$ converges weakly, as $N \to \infty$, to a Brownian motion with diffusion coefficient $D_{self}(\lambda,\bar{\rho}) = \lambda/(\lambda + \bar{\rho})$. This is the self-diffusion coefficient for the local interaction (2.4) in the diffusive scaling. We can see that, consistent with the intensity of the interaction,

$$\lim_{\lambda \to 0} \frac{\lambda}{\lambda + \bar{\rho}} = 0 \text{ and } \lim_{\lambda \to \infty} \frac{\lambda}{\lambda + \bar{\rho}} = 1 \quad , \tag{2.6}$$

the first case corresponding to the pure reflection setting, leading to a deterministic profile (constant, in equilibrium) and the second case corresponding to the noninteracting case, when $D_{self}(\lambda, \bar{\rho}) = 1$ as in (1.3). The following theorem states the main result concerning the tracer particle in nonequilibrium (see [2]).

Theorem 2.3 *If the initial density profile $\mu(0, dx)$ has a bounded initial density $\rho_0(x)$, i.e. $\mu(0, dx) = \rho_0(x)dx$ and $P^N(x_1^N(0) = x_1) = 1$ is satisfied for all $N > 0$, then the family of measures $P^N \circ (x_1^N(\cdot))^{-1}$ has a weak limit Q^{x_1} as $N \to \infty$ and Q^{x_1} is the unique solution to the martingale problem given by*

$$\mathcal{A}_t = \frac{1}{2}\left(\frac{\lambda}{\lambda + \rho(t,x)}\right)\frac{d^2}{dx^2} - \frac{1}{2}\left(\partial_x \rho(t,x)\frac{2\lambda + \rho(t,x)}{(\lambda + \rho(t,x))^2}\right)\frac{d}{dx} \tag{2.7}$$

starting at $(0, x_1)$.

The condition that the initial profile have a bounded density is not technical, as we shall explain later. Yet our interest is focused on the form of the diffusion process Q^{x_1} described in the theorem. The self-diffusion coefficient is consistent with the prediction for the equilibrium case $D_{self}(\lambda, \rho(t,x)) = \lambda/(\lambda + \rho(t,x))$. The drift depends on the spatial derivative of the density (vanishes in equilibrium) and is negative, reflecting the repulsive nature of the interaction. From a heuristic point of view, as soon as we have determined the macroscopic density $\rho(t,x)$ and $D_{self}(\lambda, \rho)$ we should be able to predict the drift term $b(t,x)$ by solving the equation

$$\rho_t = \mathcal{A}_t^* \rho \tag{2.8}$$

which yields exactly

$$b(t,x) = -\frac{1}{2}\left(\partial_x \rho(t,x)\frac{2\lambda + \rho(t,x)}{(\lambda + \rho(t,x))^2}\right) \quad . \tag{2.9}$$

For any smooth test function $\phi(x)$ on Γ we know that

$$\frac{1}{N}\sum_{i=1}^{n} \phi(x_i^N(t)) \Rightarrow \int_0^1 \phi(x)\rho(t,x)dx \tag{2.10}$$

from (2.5). In the same time, if the particles decouple in the limit, the sum $(1/N)\sum_{i=1}^{n} \phi(x_i^N(t))$ should converge to $\bar{\rho}E^{Q^\mu}[f(\omega(t)]$, where Q^μ is the law of the diffusion described in (2.7) starting from the initial distribution $\mu(0, dx)$. This implies (2.8). For the simple exclusion model this is explained in [15]. The crucial assumption we have made is that two distinct particles become independent in the limit. The next theorem and (2.12) state the propagation of chaos for our model.

Theorem 2.4 *Let $(x_1^N(\cdot), x_2^N(\cdot))$ be a pair of tagged particle processes with distinct labels. Assume that each starts at x_1, respectively x_2 on the unit circle, that is $P^N(x_1^N(0) = x_1) = 1$ and $P^N(x_2^N(0) = x_2) = 1$ and that the initial profile $\mu(0, dx)$ has bounded density $\rho_0(x)$. Let $x_1(\cdot)$ and $x_2(\cdot)$ be the two processes such that $x_1^N(\cdot) \Rightarrow x_1(\cdot)$ and $x_2^N(\cdot) \Rightarrow x_2(\cdot)$, that is there exists a measure $Q^{(x_1,x_2)}$ on $\Omega_2 = C([0,\infty), R^2)$ such that*

$$P^N \circ (x_1^N(\cdot), x_2^N(\cdot))^{-1} \Rightarrow Q^{(x_1,x_2)} \quad . \tag{2.11}$$

Then $x_1(\cdot)$ *and* $x_2(\cdot)$ *are independent with respect to* $Q^{(x_1,x_2)}$, *or equivalently* $Q^{(x_1,x_2)} = Q^{x_1} \otimes Q^{x_2}$.

The dynamics controlled by the parameter λ provides an example of hydrodynamic limit when the interaction is undetectable seen from the level of the marginals at a given time t. This is further illustrated by the fact that both the fluctuation field and the large deviations from the hydrodynamic limit behave identically as in the case of independent Brownian motions. The self-diffusion detected already in equilibrium shows that at the level of the path $\omega(\cdot) \in \Omega$ the amount of interaction is visible. In that sense we can prove the next theorem.

Theorem 2.5

$$\frac{1}{N}\sum_{i=1}^{n} \delta_{x_i^N(\cdot)} \Rightarrow \bar{\rho} Q^{\mu} \tag{2.12}$$

where Q^{μ} *is the law of tagged particle process (2.7) averaged over the initial profile* $\mu(0, dx)$.

In equilibrium this law of large numbers is an immediate consequence of Theorems 2.3 and 2.4. In nonequilibrium one cannot assume that the particles converge to their limit uniformly with respect to their label. The proof is part of a work in progress concerning the fluctuation field (2.14) which requires stronger estimates than Theorem 2.3 and Theorem 2.4. For symmetric simple exclusion processes Theorem 2.5 is proved in [12].

A way to point out the emergence of the interaction when we pass to the law of large numbers (2.12) from the hydrodynamic limit (1.2) is to think of the multicolor process. Theorem 2.3 derives the law of large numbers for the marginal at time t of the tagged particle process whereas Theorem 2.5 must take into account any finite collection of marginals at times $t_0 < t_1 < \ldots < t_m$, for any $m \in Z_+$. We may want to see what happens to just two consecutive times $0 \le s < t$. We look at the average

$$\frac{1}{N}\sum_{i=1}^{n} g(x_i^N(s), x_i^N(t)) \tag{2.13}$$

as $N \to \infty$, for an arbitrary $g(\cdot,\cdot) \in C_0^{\infty}(R^2)$. Let's consider that the particles have the same color at time $t = 0$, say blue. Also, let's pick $A \in \mathcal{B}(\Gamma)$. The particle crossing the set A at time s will change color into green. Due to the Markov property, proving the LLN for (2.13) is equivalent to finding the hydrodynamical limit of the two color process. The interaction becomes apparent because the symmetry (cancellations in the boundary conditions when averaging) is lost to the presence of the two colors. The particles will not exchange freely their labels; it matters where they come from, or, in other words, what is their history. See [10]. A large deviations principle from Q^{μ} for the empirical random measures in (2.12) for the symmetric exclusion process is obtained in [11].

In the future one would also like to give a meaning to the random field on the path space Ω of the fluctuations from the limit Q^{μ} in equilibrium

$$\zeta^N = \frac{1}{\sqrt{N}}\sum_{i=1}^{n} (\delta_{x_i^N(\cdot)} - Q^{\mu}) \quad . \tag{2.14}$$

We shall sketch the proof of Theorem 2.3 and Theorem 2.4.

3 Proof of Theorems 2.3 and 2.4

It is known (see [9]) from the definition of the process $\{x^N(t)\}_{t\geq 0}$ with the filtration $\{\mathcal{F}_t\}_{t\geq 0}$, $\mathcal{F}_t = \sigma(x^N(s) : 0 \leq s \leq t)$ that there exist $n^2 - n$ local times $\{A_N^{ij}(t)\}_{t\geq 0}$ for $i \neq j$ in the set $\{1, 2, \ldots, n\}$ such that for any $f \in \bar{C}(\Gamma^n, F)$

$$\begin{aligned} \mathcal{M}_f(t) \quad &:= \quad f(x^N(t)) - f(x^N(0)) - \frac{1}{2}\int_0^t \Delta f(x^N(s))ds - \\ &- \sum_{i\neq j}\int_0^t \left(D^{ji} f(x^N(s)) + (\lambda N)[f^{ij}(x^N(s)) - f^{ji}(x^N(s))]\right) dA_N^{ji}(s) \end{aligned} \tag{3.1}$$

is a $(P^{x^N}, \{\mathcal{F}_t\}_{t\geq 0})$ - martingale. More precisely,

$$\mathcal{M}_f(t) = \sum_{k=1}^{n}\int_0^t \partial_{x_k} f(x^N(s))d\beta_k(s) + \sum_{i\neq j}\int_0^t [f^{ij}(x^N(s)) - f^{ji}(x^N(s))]dM_N^{ji}(s)$$

where $\{\beta_k(t)\}$ $(k = 1, \ldots, n)$ is a family of independent Brownian motions and $M_N^{ij}(t)$, $M_N^{ji}(t)$ are the jump martingales corresponding to the interaction along the boundary F^{ij} such that $[M_N^{ij}(t)]^2 - (\lambda N)A_N^{ij}(t)$ is also a martingale.

A direct attempt to write the stochastic integral equation for the tagged particle $\{x_1(t)\}_{t\geq 0}$ leads to

$$x_1(t) - x_1(0) - \sum_{k\neq 1}\left[A_N^{1k}(t) - A_N^{k1}(t)\right] = \beta_1(t) \tag{3.2}$$

for the function $f_1(x) = x_1$. The term $\sum_{k\neq 1}[A_N^{1k}(t) - A_N^{k1}(t)]$ is very hard to evaluate.

Let $\nu : R \longrightarrow R$ be the periodic extension on the real line of the function $\nu(x) = x$ on $[0, 1]$. For two points x' and x'' on the unit circle $\nu(x' - x'')$ is the distance between the two in positive trigonometric sense.

In general, the process $\{((x_2^N(t) - x_1^N(t)), \ldots, (x_n^N(t) - x_1^N(t)))\}_{t\geq 0}$ is called the environment process for the tagged particle $\{x_1^N(t)\}_{t\geq 0}$. We generate a martingale with respect to the filtration $\{\mathcal{F}_t\}_{t\geq 0}$, $\mathcal{F}_t = \sigma(x^N(s) : 0 \leq s \leq t)$ and the measure P^N denoted by

$$z_1^N(t) := x_1^N(t) + \frac{1}{\lambda + \bar{\rho}} \cdot \frac{1}{N}\sum_{k\neq 1} \nu(x_k^N(t) - x_1^N(t)) \tag{3.3}$$

as a linear combination of the two. There are several steps to follow.

3.1 Step 1: The asymptotic average collision time per particle. The quadratic variation of the martingale $z_1^N(t)$ is

$$\sigma^2(N, \lambda)t + \frac{1}{(\lambda + \bar{\rho})^2}\left(\frac{\lambda}{N}\right)\sum_{k\neq 1}\left(A_N^{1k}(t) + A_N^{k1}(t)\right) \tag{3.4}$$

where

$$\sigma^2(N, \lambda) = \left[1 - \frac{n-1}{N(\lambda + \bar{\rho})}\right]^2 + \frac{1}{(\lambda + \bar{\rho})^2} \cdot \frac{(n-1)}{N^2} \quad .$$

The average interaction local time per particle

$$A_N^1(t) := \frac{1}{N}\sum_{k\neq 1}\Big(A_N^{1k}(t)+A_N^{k1}(t)\Big) \tag{3.5}$$

is the main quantity to evaluate in the limit.

Theorem 3.1 *For any initial profile $\mu(0,dx)$*
1) the average interaction local time per particle $\{A_N^1(\cdot)\}_N$ is tight and
2) $dA_N^1(t)$ is asymptotically equal to $\rho(t,x_1^N(t))dt$, i.e. $\forall\, t\geq 0$

$$\lim_{N\to\infty} E^N\left|A_N^1(t)-\int_0^t \rho(s,x_1^N(s))ds\right| = 0 \quad . \tag{3.6}$$

The hardest part of the proof of theorem 2.3 is to establish (3.6).

3.2 Step 2: The tightness. Theorem 3.1 shows that $z_1^N(t)$ is tight. At time $t=0$ the martingale $z_1^N(t)$ takes values in a bounded set $[0,2]$. Doob's inequality and the asymptotic estimate (3.6) take care of the rest. The idea underlying the proof of tightness for the tagged particle process $x_1^N(t)$ is that for $s<t$ the quantity

$$|z_1^N(t)-z_1^N(s)| \tag{3.7}$$

$$= \left|x_1^N(t)-x_1^N(s)+\frac{1}{\lambda+\bar\rho}\left(\frac{1}{N}\sum_{k\neq 1}(\nu(x_k^N(t)-x_1^N(t))-\nu(x_k^N(s)-x_1^N(s)))\right)\right|$$

must be large if $|x_1^N(t)-x_1^N(s)|$ is large. This will show that if $z_1^N(t)$ is tight, then $x_1^N(t)$ is tight as well.

3.3 Step 3: The connection between the martingale $z_1^N(t)$ and $x_1^N(t)$. It is important to remark that both Step 1 and Step 2 of the proof make no use of the assumption that $\mu(0,dx)=\rho_0(x)dx$ for $\rho_0(x)$ bounded. Let's denote the path space $\{\omega(\cdot):\omega\in C([0,T])\}$ by Ω. The mapping $\Theta:\Omega\longrightarrow\Omega$

$$\Theta(\omega)(t) := \omega(t)+\frac{1}{\lambda+\bar\rho}\int_0^1 \nu(y-\omega(t))\rho(t,y)dy \tag{3.8}$$

is one-to-one and onto if $\rho_0(x)$ is bounded. We want to have

$$\lim_N E^{P^N}\left|\Big(x_1^N(t)+\frac{1}{\lambda+\bar\rho}\frac{1}{N}\sum_{k\neq 1}\nu(x_k^N(t)-x_1^N(t))\Big)\right. \tag{3.9}$$

$$\left.-\Big(x_1^N(t)+\frac{1}{\lambda+\bar\rho}\int_0^1 \nu(y-x_1^N(t))\rho(t,y)dy\Big)\right| = 0 \quad . \tag{3.10}$$

This is done by extending the hydrodynamic limit (1.2) to functions with a finite number of discontinuities like $\nu(x)$ and introducing $x_1^N(t)$ as a parameter. There are many details to fill in, but in the end (3.9) implies that, if $x_1(\cdot)$ is a limit point for the tight processes $\{x_1^N(\cdot)\}_{N>0}$, as well as $z_1(\cdot)$ is a limit point for $\{z_1^N(\cdot)\}_{N>0}$, then

$$z_1(\cdot) = \Theta(x_1(\cdot)) \quad \text{a.s.} \ . \tag{3.11}$$

For every $t\in[0,T]$ the function

$$z = F(t,x) = x+\frac{1}{\lambda+\bar\rho}\int_0^1 \nu(y-x)\rho(t,y)dy$$

has an inverse $G(t,z)$. With this notation, the asymptotic behavior of $A_N^1(t)$ proves that the martingale $z_1^N(\cdot)$ converges to a diffusion with generator

$$\mathcal{A}_t^z = \frac{1}{2}\frac{\lambda(\lambda+\rho(t,x))}{(\lambda+\bar{\rho})^2}\frac{d^2}{dz^2} \quad . \tag{3.12}$$

However, we know $z_1(\cdot)$ from (3.12). It is sufficient to show that if two processes supported on Ω are related as in (3.11) then the tagged particle process $x_1(\cdot)$ is a diffusion with generator (2.7). This fact is shown in the next lemma.

Lemma 3.2 *We assume that the martingale problem is well posed for the pair $(a(t,y),b(t,y))$, i.e. for any $(t,y)\in[0,T]\times R$ there is a measure $P^{(s,y)}$ on the path space $\Omega = C([0,T],R)$ such that if $y(\cdot)$ denotes an element of Ω and*

$$\mathcal{L}_t := \frac{1}{2}a(t,y)\frac{d^2}{dy^2} + b(t,y)\frac{d}{dy}$$

then

1)$P^{(s,y)}(\{y(s)=y\}) = 1$ and

2) $\forall f(\cdot,\cdot)\in C_0^\infty([0,T],R)$ the expression

$$f(t,y(t)) - f(s,y(s)) - \int_s^t (\partial_u + \mathcal{L}_u)f(u,y(u))du$$

is a $(P^{(s,y)},\mathcal{F}_t)$-martingale, where $\mathcal{F}_t = \sigma(\omega(s) : 0\le s\le t)$.

Suppose $\Theta : [0,T]\times R \to R$ is a C^2 mapping such that

1) $(t,\Theta(t,x)) = (t,y)$ and

2) $0 < c_1 \le \partial_x\Theta(t,x) \le c_2 < \infty$ for any (t,x). Then $x\to\Theta(t,x)$ has an inverse $y\to\Psi(t,y)$ for any fixed $t\ge 0$. If we define a mapping on the path space $\Xi : \Omega\to\Omega$ by $[\Xi(y)](t) := \Psi(t,y(t)) = x(t)$, then $\hat{P}^{(s,x)} := P^{(s,\Theta(s,x))}\circ\Xi^{-1}$ solves the martingale problem $(\hat{a}(t,x),\hat{b}(t,x))$ with

$$\hat{a}(t,x) = [a\cdot(\partial_y\Psi)^2]\circ(t,\Theta(t,x)) \tag{3.13}$$

and

$$\hat{b}(t,x) := \left[(\partial_t\Psi) + \frac{1}{2}a\cdot(\partial_{yy}\Psi) + b\cdot(\partial_y\Psi)\right]\circ(t,\Theta(t,x)) \quad . \tag{3.14}$$

4 Relaxing the initial conditions

The mapping Θ defined in (3.8) is not well defined for singular measures on the circle. The limit

$$\lim_{t\to 0}(\Theta\omega)(t) = \lim_{t\to 0}\left[\omega(t) + \frac{1}{\lambda+\bar{\rho}}\int_{[0,1]}\nu(y-\omega(t))\rho(t,y)dy\right] \tag{4.1}$$

does not exist for paths starting at $\omega(0) = x_1$ if the $\mu(0,\{x_1\}) > 0$. The tagged particle moves both due to the Brownian character of the evolution and the interaction with the environment process. For $t>0$ the density $\rho(t,x)$ is smooth for an arbitrary initial profile, which, by Theorem 3.1, grants its uniqueness for $t' > t$. This suggests that the uniqueness of the whole process must be exclusively dependent on the outlook of the environment at time $t=0$, that is on $((x_2^N(0)-x_1^N(0)),\ldots,(x_n^N(0)-x_1^N(0)))$ at time $t=0$. A tracer particle will be pushed away from the direction which carries more weight in the interaction, that is, where there is a larger number of particles per volume. Hence we must know the amount of mass on both sides of x_1, even though we already know the total mass piling up at this point, namely $\mu(0,\{x_1\})$. A way of making these considerations precise is the assumption stated in equation (4.4).

We define two periodic functions of period one on R. On the unit interval $[0,1)$ they are equal to

$$\phi_\epsilon(x) = x, \text{ if } x \in [0, 1-\epsilon], \quad x-1, \text{ if } x \in [1-\frac{\epsilon}{2}, 1] \quad \text{and smooth on } [0,1) \tag{4.2}$$

and

$$g_\epsilon(x) = \nu(x) - \phi_\epsilon(x) \quad . \tag{4.3}$$

If $\mu(\{x_1\}) := \pi(x_1) > 0$ we assume that there is a number $\pi_-(x_1)$ in the interval $[0, \pi(x_1)]$ (clearly $\pi_-(x_1) = 0$ if $\mu(\{x_1\}) = 0$) such that

$$\lim_{\epsilon\to 0} \limsup_{N\to\infty} E^N \left| \frac{1}{N} \sum_{k\neq 1} g_\epsilon(x_k^N - x_1^N) - \pi_-(x_1) \right| = 0 \quad . \tag{4.4}$$

For any x_0 let $\Omega_{x_0} = \{\omega : \omega \in \Omega \text{ and } \omega(0) = x_0\}$. The mapping Θ (3.8) will be altered at the starting point. We write the function $F(t,x)$

$$F(t,x) = x + \frac{1}{\lambda + \bar\rho} \int_0^1 \nu(y-x)\rho(t,y)dy \tag{4.5}$$

for $t > 0$ and equal to the constant

$$z_1 := x_1 + \frac{1}{\lambda+\bar\rho} \int_{[0,1]\setminus\{x_1\}} \nu(y-x_1)\mu(dx) + \frac{1}{\lambda+\bar\rho}\pi_-(x_1) \tag{4.6}$$

for $(0, x_1)$. This makes

$$\Theta(\omega)(t) = F(t, \omega(t)) \text{ for } t > 0 \text{ and } \Theta(\omega)(0) = z_1$$

a mapping from Ω_{x_1} into Ω_{z_1}. The point z_1 will be the "adjusted" starting point for (3.3). The main result concerning uniqueness is the following theorem, from [3]. The infinitesimal generator $\mathcal{A}_t$ of the process has been defined in (2.7).

Theorem 4.1 *Under (1.1) and (4.4), provided that $P^N(x_1^N(0) = x_1) = 1$ for all $N > 0$, we have*

1) If $\mu(dx)$ is continuous at x_1, i.e. $\mu(\{x_1\}) = 0$ then the tagged particle process starting at x_1 is unique.

2) If $\mu(\{x_1\}) = \pi(x_1) > 0$, then the family of limit points of the tight family of processes $\{P^N \circ (x_1^N(\cdot))^{-1}\}_N$ is infinite and there is a unique such limit for each value of $\pi_-(x_1) \in [0, \pi(x_1)]$ denoted for simplicity by Q^{x_1}.

3) Q^{x_1} can be characterized as the unique measure on Ω_{x_1} with the properties

(i) For any smooth $f(t,x)$ with $supp(f) \subseteq (0,\infty)\times R$ the expression

$$f(t, x_1(t)) - f(0, x_1(0)) - \int_0^t (\partial_u f + \mathcal{A}_u f)(u, x_1(u))du$$

is a $(Q^{x_1}, \{\mathcal{F}_t\}_{t\geq 0})$ - martingale and

(ii) $\quad Q^{x_1}(\{\omega \in \Omega_{x_1} : \Theta\omega \in \Omega_{z_1}\}) = 1.$

References

[1] Chang, C. C.; Yau, H-T. Fluctuations of one-dimensional Ginzburg-Landau models in nonequilibrium. *Comm. Math. Phys.* 145 (1992), no. 2, 209-234.

[2] Grigorescu, I. (1997) Self-diffusion for Brownian motions with local interaction. *Annals Prob.*, to appear.

[3] Grigorescu, I. Uniqueness of the tagged particle process in a system with local interactions. *Annals Prob.* , to appear.

[4] Guo, M. Z. (1984) Limit Theorems for Interacting Particle Systems. *Ph. D. thesis*, New York University.

[5] Guo, Mao Zheng; Papanicolaou, George. (1984) Bulk diffusion for interacting Brownian particles. *Statistical physics and dynamical systems* (Koszeg,1984), 41-48, Progr. Phys., 10, Birkhauser Boston, Boston, Mass., 1985.

[6] Guo, Mao Zheng; Papanicolaou, George. (1985) Self-diffusion of interacting Brownian particles. *Probabilistic methods in mathematical physics*(Katata/Kyoto, 1985), 113-151

[7] Kipnis, C.; Olla, S. Large deviations from the hydrodynamical limit for a system of independent Brownian particles. *Stochastics Stochastic reports.* 33 (1990), no. 1-2, 17-25.

[8] Kipnis, Claude; Varadhan, S. R. S. Central limit theorems for additive functionals of reversible Markov chains and applications. Colloquium in honor of Laurent Schwartz, Vol. 2 (Palaiseau, 1983). Astrisque No. 132 (1985), 65–70.

[9] Ikeda, N. & Watanabe, S. (1981) Stochastic Differential Equations and Diffusion Processes. *North-Holland, Amsterdam (Kodansha Ltd., Tokyo)*

[10] Quastel, J. Diffusion of Color in the simple exclusion process. *Comm. Pure. Appl. Math.* 45 (1992), no. 6, 623-679

[11] Quastel, J.; Rezakhanlou, F.; Varadhan, S. R. S. Large deviations for the symmetric simple exclusion process in dimensions $d \geq 3$. *Probab. Theory Related Fields* 113 (1999), no. 1, 1–84.

[12] Rezakhanlou, F. Propagation of chaos for symmetric simple exclusions. *Comm. Math. Phys.* 140 (1991), no.3, 417-448.

[13] Spohn, H. (1986) Equilibrium fluctuations for Interacting Brownian Particles. *Comm. Math. Phys. 103* No. 1, 1-33.

[14] Spohn, H. Tracer diffusion in lattice gases. *J. Stat. Phys.* 59 (1990), no. 5-6, 1227-1239.

[15] Varadhan, S. R. S. The complex story of simple exclusion. It's stochastic calculus and probability theory, 385–400, Springer, Tokyo, 1996.

[16] Varadhan, S. R. S. Self-diffusion of a tagged particle in equilibrium for asymmetric mean zero random walk with simple exclusion. *Ann. Inst. H. Poincar Probab. Statist.* 31 (1995), no. 1, 273–285.

[17] Varadhan, S. R. S. Regularity of self-diffusion coefficient. The Dynkin Festschrift, 387–397, Progr. Probab., 34, Birkhuser Boston, Boston, MA, 1994.

[18] Varadhan, S. R. S. Scaling limits for interacting diffusions. *Comm. Math. Phys.* 135 (1991), no. 2, 313–353.

Fields Institute Communications
Volume **27**, 2000

Stellar Dynamics and Plasma Physics with Corrected Potentials: Vlasov, Manev, Boltzmann, Smoluchowski

Reinhard Illner
Department of Mathematics and Statistics
University of Victoria, P.O. Box 3045
Victoria, B.C. V8W 3P4, Canada
rillner@math.uvic.ca

Abstract. We summarize the results from several recent papers in which the effects of a particle interaction via an (attractive or repulsive) potential $\pm 1/r^2$ on the associated Vlasov, Boltzmann, or, most generally, Vlasov-Boltzmann-Smoluchowski equations were discussed. The potential is considered in isolation; it occurs in the literature as a correction ϵ/r^2 to the Coulomb potential, and the associated law is known as Manev potential.

1 Motivation and background

This paper will summarize the results from Refs. [1, 2, 5]. We are concerned with particle clouds which interact via a pair potential $u(r) = \pm 1/r^2$. For $u(r) = -1/r^2$, forces are attractive, and the resulting kinetic equations are of stellar dynamic type; for $u(r) = 1/r^2$, the repulsive case, the kinetic equations are Vlasov equations for plasmas (charged particle systems). This kind of potential has been studied since Newton [11], who was the first to describe some of the interesting effects arising in the central force problem. Manev [7]-[10] studied u as a correction to the Newtonian potential; specifically, he considered

$$-\frac{\gamma}{r} - \frac{\epsilon}{r^2},$$

where γ is assumed to be the universal gravitational constant and $\epsilon = 3\gamma^2/c^2$ is thought of as a quasirelativistic correction (for a justification of this interpretation see [5]). This choice of constants predicts the qualitatively correct precession of the perihelion of Mercury. That potentials of this kind would lead to precessional ellipses as orbits was already known to Newton [11].

It is an interesting question, at least from a mathematical point of view, to investigate what happens to kinetic theory if the Manev correction (as we will from now on call it) is incorporated into the kinetic equations. This is exactly was was

2000 *Mathematics Subject Classification.* Primary 28C22; Secondary 82C40.

done in Refs. [1, 2, 5]. The major milestones, which we will revisit in the current paper, are the following:

1. The initial value problem for a Vlasov equation with self-consistent interparticle potential $1/r^2$ possesses unique local solutions for sufficiently smooth initial data. Nonnegativity and the usual conservation laws (mass, momentum, energy, angular momentum) apply.
2. There are initial values for which global existence does not hold in the attractive case. This is proved indirectly from a relationship linking the moment of inertia with the (invariant) total energy. If the latter is negative, an upper estimate on the maximal existence time is obtained.
3. The Vlasov equation with the self-consistent potential $1/r^2$ possesses an additional fundamental invariant (referred to as "projective invariant") in addition to the usual scalings of time, space, velocity, and Galilei invariance.
4. If particles interact via an (attractive or repulsive) potential $1/r^2$, the addition of Boltzmann collision terms is suggested by dimensional analysis.
5. In the attractive case, inclusion of a Smoluchowski coagulation term is justified by the analysis of pair interactions. Specifically, there are particle encounters which lead to coagulation in the sense that particles will coagulate for small enough collision parameters.

Many of these results are, as indicated, specific for attractive or repulsive forces; they remain valid if the Manev correction is just added to the Newtonian (or Coulomb) potential. As already stated, we will here consider the Manev correction in isolation; we will also concentrate on the attractive case, which displays the most interesting mathematical features.

2 Equations, invariants, non-global existence

We focus first on the equation in which a Vlasov-type force with an attractive interparticle potential $-\epsilon/r^2$ is responsible for the forces. $f = f(t, x, v)$ denotes the density function at time t in $\mathbb{R}^6$, and $\rho(t, x) = \int f(t, x, v)\, dv$ is the spatial density. The potential is then

$$U[\rho](t, x) = -\epsilon \int \frac{\rho(t, y)}{|x - y|^2}\, dy$$

and the force at location x at time t is

$$\begin{aligned} F(t, x) = -\nabla_x U[\rho](t, x) &= -\epsilon \int \frac{1}{|x - y|^2} \nabla_y \rho(t, y)\, dy \\ &= -2\epsilon \int \frac{x - y}{|x - y|^4} \rho(t, y)\, dy. \end{aligned} \tag{2.1}$$

Note that the last integral on the right must be interpreted as a Cauchy principal value; the integration domain is always all space, so the integral is defined, e.g., if ρ is at least Hölder continuous. The evolution equation for f is

$$\partial_t f + v \cdot \nabla_x f + F \cdot \nabla_v f = 0, \tag{2.2}$$

with F given as above. We call (2.2) the stellar dynamic equation associated with the potential $-\epsilon/r^2$. It is nonlinear because F depends implicitly on ρ, and ρ is given in terms of f. A major difference with respect to the classical stellar dynamic equation (associated with the potential $-\gamma/r$) is that even the definition of F requires that ρ posesses some smoothness (whereas in the classical case boundedness and integrability are sufficient).

Assuming that we have a sufficiently smooth solution, one readily proves, using the method of characteristics, that for all $1 \leq p \leq \infty$

$$\|f(t,\cdot)\|_{L^p} = \|f_0\|_{L^p} \tag{2.3}$$

(which includes conservation of mass) and that

$$\frac{d}{dt}\iint vf(t,x,v)\,dv = 0 \tag{2.4}$$

$$\frac{d}{dt}\left[\iint v^2 f\,dv\,dx - \epsilon \iint \frac{\rho(t,x)\rho(t,y)}{|x-y|^2}\,dx\,dy\right] = 0 \tag{2.5}$$

(momentum and energy conservation). Conservation of angular momentum also holds.

The proof that the initial value problem is well-posed is actually quite involved and was given in Ref. [5]. One has to work in spaces of functions with Hölder continuous derivatives; in particular, the initial values have to have this degree of regularity. The force field given in (2.1) is undefined at discontinuities of ρ. The model therefore loses its validity if such discontinuities form, or if ρ develops strong enough singularities.

It is remarkably easy to see that this must actually happen. The method is based on a standard calculation used in the theory of N-body problems, the theory of the nonlinear Schrödinger equation, and also in classical stellar dynamics.

Consider the functional $I(t) = \int x^2 \rho(t,x)\,dx$ (twice the moment of inertia of the system). A straightforward calculation shows that

$$\frac{d^2}{dt^2} I(t) = E(t) = E(0) \tag{2.6}$$

where $E(t)$ denotes the (time-invariant) energy, given by the expression inside the bracket in (2.5). If $E(0)$ is negative (meaning that there is not a lot of "temperature" in the system), it follows that $I(t)$ must eventually become negative, contradicting the fact that $\rho \geq 0$. Hence the solution must cease to exist at some earlier time. The time when $I(t)$ becomes negative is easily calculated by integrating (2.6) twice. For details, see [1].

3 Projective invariance

More than 100 years ago Sophus Lie classified the group invariants of the free flow characteristic equations $\frac{d^2}{dt^2}x(t) = 0$ (or $\dot{x} = v, \quad \dot{v} = 0$). There are 4 of those:

$$\begin{array}{ll} \text{Shift:} & t' = t + \alpha_1, \quad x' = x \\ \text{Scale:} & t' = e^{\alpha_2} t, \quad x' = x \\ \text{Galilei:} & t' = t, \quad x' = x + \alpha_3 t \\ \text{Projective:} & t' = \dfrac{t}{1-\alpha_4 t}, \quad x' = \dfrac{x}{1-\alpha_4 t}. \end{array}$$

If x is one-dimensional, there are 4 more by interchanging x and t. The constants $\alpha_1, \ldots, \alpha_4$ are free parameters. We refer to [3] for a detailed discussion of these invariants.

The first three invariants remain valid for the $N-$body problem (i.e., $x \in \mathbb{R}^{3N}$) in the presence of a conservative interparticle potential, like the Newtonian potential. However, the projective invariance is generally lost at this level, with one exception: For interparticle potentials $\pm\epsilon/r^2$, projective invariance remains valid, and remains valid even for the Vlasov equation under consideration here. Specifically, the following theorem holds.

Theorem 3.1 *Let f(t,x,v) be a solution of (2.2), where the force is given by (2.1) but can be attractive or repulsive. Suppose that f exists on a time interval $[0, t_0)$. Let $a > 0$ and set*

$$\tau = \frac{t}{1+at}, \quad y = \frac{x}{1+at}, \quad w = v(1+at) - ax.$$

Then $F(t, y, w) := f(t, x, v)$ solves Eqn. (2.2) with respect to τ, y, w on an interval $\tau \in \left[0, \frac{t_0}{1+at_0}\right]$, and for the initial value $F(0, y, w) = f_0(y, w + ay)$.

This projective invariance of the equation has interesting consequences: it allows to transform solutions into other solutions and even to produce some global solutions from local solutions. Note that the transformation $(t, x, v) \to (\tau, y, w)$ is invertible; we can write

$$t = \frac{\tau}{1-a\tau}, \quad x = \frac{y}{1-a\tau} \quad v = w(1-a\tau) + ay,$$

and we see that if $F(\tau, y, w)$ solves the equation on $[0, \tau_0]$, then $f(t, x, v) = F(\tau, y, w)$ assumes the initial value $F_0(x, v - ax)$ and exists on the time interval $\left[0, \frac{\tau_0}{1-a\tau_0}\right]$ if $a < 1/\tau_0$, and on $[0, \infty)$ if $a \geq 1/\tau_0$. It is transparent how the transformation changes the data to new data with different angular momentum.

The proof of the theorem is elementary and uses only the chain rule and the representation of the force field given by (2.1).

Assume that $f = f(x, v)$ is a nonnegative steady solution; the existence of such solutions for the attractive case has recently been shown by Steacy [12]. We can then apply the above transformation and get that $f\left(\frac{x}{1-at}, v(1-at) + ax\right)$ is also a solution, assuming the initial value $f(x, v+ax)$ and ceasing to exist at time $t = 1/a$.

4 Why a Boltzmann collision term should be added

We now discuss a scaling argument first presented in [1] which will suggest that the equation discussed so far is incomplete. For this section, we consider a more general repulsive interparticle potential $u(r) = \alpha/r^n$, where $\alpha > 0$ is a constant and $1 < n \leq 2$. In order to accurately describe close encounters (collisions) between particles, we add a Boltzmann collision term $Q(f, f)$ to the right hand side of the equation. The differential cross-section associated with the potential is

$$\sigma(|v|, \theta) = \left(\frac{\alpha}{m|v|^2}\right)^{2/n} g_n(\cos\theta),$$

where $|v|$ is the relative particle speed, θ is the scattering angle, m is the particle mass and g_n is a function such that $\int_{-1}^{1}(1-x)g_n(x)\, dx < \infty$. The Vlasov-Boltzmann equation is then

$$\partial_t f + v \cdot \nabla_x f + \frac{F}{m} \cdot \nabla_v f = Q(f, f) \tag{4.1}$$

with

$$F = -\nabla_x \int \frac{\alpha}{|x-y|^n} \rho(t, y)\, dy$$

and

$$Q(f,f) = \iint |v - v_*|\,\sigma(|v - v_*|,\theta)\,\{f'f'_* - ff_*\}\sin\theta\,d\theta\,d\varphi\,dv_*,$$

with the usual notation in the Boltzmann collision term: $f_* = f(t,x,v_*)$, $f'_* = f(t,x,v'_*)$, etc., where v, v_* are pre- and v', v'_* are post-collisional velocities respectively.

The term involving F and the term $Q(f,f)$ in (4.1) represent effects of long-range and short-range forces respectively. To decide which one dominates in the particle system under consideration, we pass to a dimensionless representation: Assume that x_0, v_0, t_0 (with $x_0 = v_0 t_0$) and ρ_0 are typical scales for the system under consideration. We then define new, dimensionless independent and dependent variables via

$$\tilde{x} = \frac{x}{x_0}, \quad \tilde{v} = \frac{v}{v_0}, \quad \tilde{t} = \frac{t}{t_0} \quad \text{and} \quad \tilde{f}(\tilde{t},\tilde{x},\tilde{v}) = \frac{v_0^3}{\rho_0} f(t,x,v).$$

It is tedious but easy to rewrite Eqn. (4.1) in terms of the new variables. After this is done and we drop all the tildas, the equation reads

$$\partial_t f + v\cdot\nabla_x f - C_V \nabla_x \int \frac{\rho(t,y)}{|x-y|^n}\,dy \cdot \nabla_v f = C_B Q(f,f) \tag{4.2}$$

The constants C_V and C_B are

$$\begin{aligned} C_V &= \rho_0 \left(\frac{\alpha}{m v_0^2}\right) x_0^{3-n} \\ C_B &= \rho_0 \left(\frac{\alpha}{m v_0^2}\right)^{2/n} x_0. \end{aligned}$$

We define $r_* = \left(\frac{\alpha}{m v_0^2}\right)^{1/n}$. Note that r_* is essentially the square root of the differential cross-section and has therefore the dimension of length. In fact, it can be interpreted as a distance over which typical particles strongly correlate. Furthermore, we define the quotient

$$\lambda = \frac{C_V}{C_B} = \left(\frac{\alpha}{m v_0^2}\right)^{\frac{n-2}{n}} x_0^{2-n} \tag{4.3}$$

If we interpret $\rho_0^{-1/3}$ as a measure for the typical distance between two particles in the system, the validity of the kinetic description requires that $r_* \ll \rho_0^{-1/3} \ll x_0$. Hence the quotient $\beta := r_*/x_0$ will satisfy $\beta \ll 1$, and we observe that

$$\lambda = \beta^{n-2} \approx \begin{cases} 0 & \text{if} \quad n > 2 \\ 1 & \text{if} \quad n = 2 \\ \infty & \text{if} \quad n < 2. \end{cases}$$

Hence if $n > 2$, the Boltzmann collision term dominates over the Vlasov force term (note, however, that this is a purely formal observation as we have made no attempt to define the Vlasov potential for this case— additional regularity assumptions on the density are needed to do so); if $n < 2$, the Vlasov force term dominates. It is exactly for $n = 2$, i.e., the case we are discussing here in detail, that both terms carry similar weight.

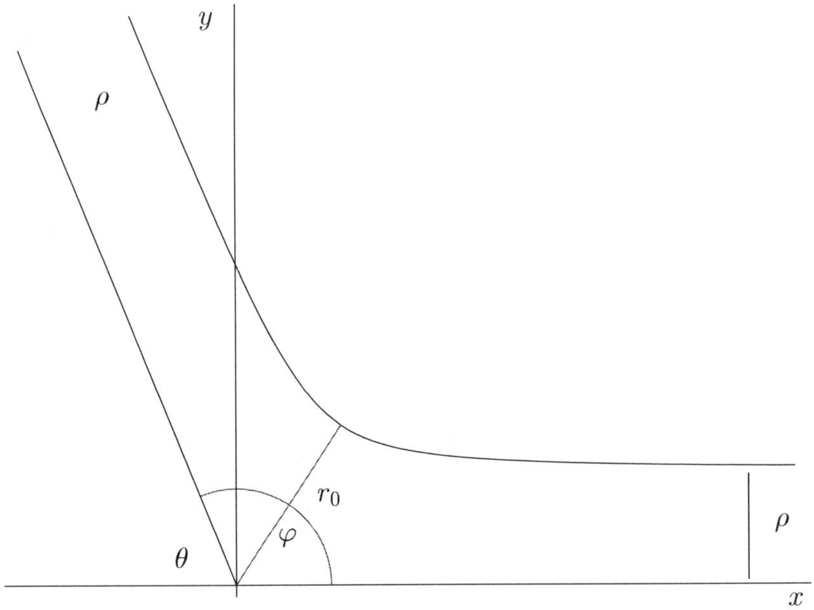

Figure 1 Scattering with a repulsive potential.

5 The scattering problem for attractive potentials

The previous section suggests that Boltzmann collision terms should also be taken seriously when particles interact via an attractive pair potential $-1/r^2$. In fact, two new phenomena arise which require careful attention: One has to address the phenomenon that the relationship between collision parameter and scattering angle is not a one-to-one function for attractive forces, and, for the specific potential $-1/r^2$, that there is the possibility of coagulation. The latter gives rise to yet another term on the right hand side of the equation, namely, a coagulation term of Smoluchowski type.

To start, let us revisit the scattering formulas for the classical repulsive situation. Assume that the interparticle potential $U(r)$ is repulsive and that two particles move at relative speed u and with impact parameter $\rho > 0$, as depicted in Figure 1. r_0 denotes the distance of closest approach between the particles. The angle φ depends on ρ, u and r_0 like

$$\varphi = 2\rho \int_{r_0}^{\infty} \frac{1}{r^2\sqrt{1 - \frac{2U(r)}{mu^2} - \left(\frac{\rho}{r}\right)^2}}\, dr. \tag{5.1}$$

Here, m is the particle mass. A derivation of (5.1) is a straightforward application of the conservation of energy and momentum and can be found in textbooks on classical mechanics, see, e.g., [6]. r_0 is determined as the distance for which the expression under the square root is zero.

Note that $\varphi \in [0, \infty]$ with $\varphi(0) = 0, \quad \varphi(\infty) = \pi$. The function $\rho \to \varphi$ is 1-1 and invertible. Setting $\theta = \pi - \varphi$, we define $\tilde{\rho}(\theta) = \rho(\pi - \theta)$. The differential cross-section to be used in the Boltzmann collision kernel is

$$\sigma(u,\theta) = \frac{\tilde{\rho}(\theta)}{\sin\theta}\left|\frac{d\tilde{\rho}(\theta)}{d\theta}\right|. \tag{5.2}$$

As is known [4], the function $|u|\sigma(u,\theta)$ is the appropriate kernel for the Boltzmann collision integral associated with the potential $U(r)$ under consideration.

Let us first study a power-potential

$$U(r) = -\frac{\alpha}{r^\gamma}, \quad \alpha > 0, \quad 1 \le \gamma < 2.$$

Formula (5.1) becomes

$$\varphi = 2\rho\int_{r_0}^{\infty}\frac{dr}{r^2\sqrt{1+\left(\frac{s}{r}\right)^\gamma - \left(\frac{\rho}{r}\right)^2}}, \quad \text{with} \quad s^\gamma = \frac{2\alpha}{mu^2}, \tag{5.3}$$

or

$$\varphi = \varphi(z) = 2\int_0^{x_0}\frac{dx}{\sqrt{1+zx^\gamma - x^2}},$$

with $z = \frac{2\alpha}{mu^2\rho^\gamma}$ and $1 + zx_0^\gamma - x_0^2 = 0$.

Consider the special powers $\gamma = 2 - 1/n$, $n = 1, 2, \ldots$, then the substitution $x = (zy)^n$ leads to

$$\varphi(z) = 2n\int_0^{y_0} dy\left[z^{-2n}y^{2-2n} + y(1-y)\right]^{-1/2}$$

with $y_0(1-y_0) + y^{2-2n}z^{-2n} = 0$.

If $z \to \infty$, i.e., $\rho \to 0$, then

$$\varphi(z) \to \varphi_{\max} = 2n\int_0^1\frac{dy}{\sqrt{y(1-y)}} = 2n\pi. \tag{5.4}$$

Hence simple trajectories (i.e., $\pi < \varphi < 2\pi$) occur only for $n = 1$ (the Newtonian potential), while in the general cases $n = 2, 3, \ldots$ we get spiral trajectories with $n-1$ loops. See Figure 2 for an example.

To compute the scattering angle $\theta \in [0,\pi]$ between the velocities $\vec{u}_-$ and $\vec{u}_+$, we observe the following elementary geometric relationship between $\varphi \in (\pi,\infty)$ and $\theta \in [0,\pi]$ (see Figure 2):

$$\text{If} \quad (2k-1)\pi \le \varphi \le 2k\pi, \quad \text{then} \quad \theta = \varphi - (2k-1)\pi, \tag{5.5}$$

$$\text{and if} \quad 2k\pi \le \varphi \le (2k+1)\pi, \quad \text{then} \quad \theta = (2k+1)\pi - \varphi. \tag{5.6}$$

The formulas (5.1), (5.5) and (5.6) define the scattering angle θ uniquely as a function of the impact parameter $\rho \in (0,\infty)$.

We next consider the scattering problem for the potential

$$U(r) = -\frac{\epsilon}{r^2}, \quad \epsilon > 0. \tag{5.7}$$

For this case, the integral (5.1) gives the simple explicit formula

$$\varphi = \pi\left[1 - \left(\frac{\rho_*}{\rho}\right)^2\right]^{-1/2}, \quad \text{with} \quad \rho_*^2 = \frac{2\epsilon}{mu^2} \tag{5.8}$$

for $\rho > \rho_*$. Note that $\varphi \to \infty$ as ρ decreases from ∞ to ρ_*.

If $\rho < \rho_*$, then the particle "falls into the center $r = 0$". To analyse this in more detail, recall that the scattering problem appears initially from a reduction of the two body problem. Consider two particles with masses m_i, positions $\vec{x}_i$

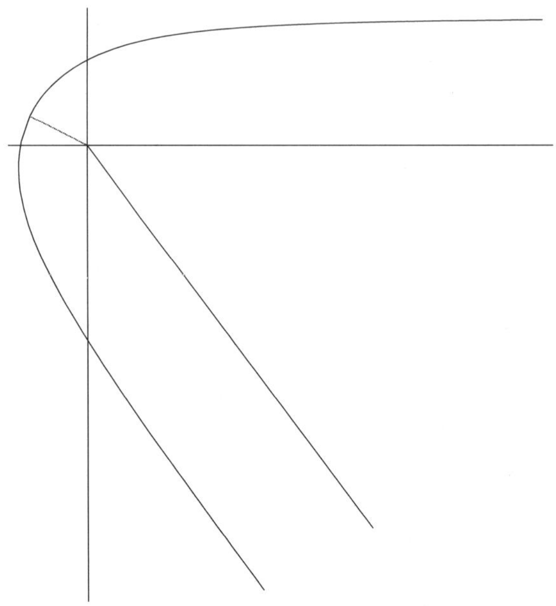

Figure 2 $\pi < \varphi \leq 2\pi$, and $\theta = \varphi - \pi$.

and velocities $\vec{v}_i$ $(i = 1, 2)$, interacting via the potential $U(|\vec{x}_1 - \vec{x}_2|)$. A standard transformation to the center-mass frame of reference,

$$\begin{aligned}\vec{X} &= \frac{m_1\vec{x}_1 + m_2\vec{x}_2}{m_1 + m_2}, \quad \vec{x} = \vec{x}_1 - \vec{x}_2, \\ \vec{V} &= \frac{m_1\vec{v}_1 + m_2\vec{v}_2}{m_1 + m_2}, \quad \vec{u} = \vec{v}_1 - \vec{v}_2\end{aligned} \tag{5.9}$$

reduces the two-body problem to the problem of one body (with mass $m = \frac{m_1 m_2}{m_1 + m_2}$, position $\vec{x}$ and velocity $\vec{u}$) in the central field $U(|\vec{x}|)$. For the potential (5.7) a global (in time) solution of the two-body problem does not exist for sufficiently small ($\rho < \rho_*$) relative angular momentum. For given initial conditions

$$\begin{aligned}\vec{x}(0) &= \vec{x}_0, \quad \vec{u}(0) = \vec{u}_0, \quad \vec{x}_0 \cdot \vec{u}_0 < 0, \quad E = \frac{m\vec{u}_0^2}{2} - \frac{\epsilon}{|\vec{x}_0|^2} < 0, \\ M^2 &:= m^2[x_0^2 u_0^2 - (\vec{x}_0 \cdot \vec{u}_0)^2] < 2\epsilon m,\end{aligned}$$

one can prove that there exists a time instant $t_0 \in (0, \infty)$ such that

$$\vec{x}(t) \to 0, \quad |\vec{u}(t)| \to \infty \quad \text{as} \quad t \nearrow t_0.$$

The question arises how to continue the solution of the two-body problem for $t > t_0$. A natural way to do so is to assume that the two particles simply coagulate at $t = t_0$, i.e., they form a larger particle with mass $M = m_1 + m_2$ and velocity $\vec{V}$. This guarantees momentum conservation, while there is in general a loss of energy. This corresponds to the implicit assumption that the new "large" particle possesses internal energy which, however, is irrelevant for our purposes.

Summarizing, we describe collisions between two particles with masses m_i, $i = 1,2$ and velocities $\vec{v}_i$, $i = 1,2$ before the collision and interacting with the potential (5.7) as follows. If the impact parameter ρ satisfies $\rho > \rho_*$ (see (5.8)), then an elastic scattering with scattering angle $\theta \in [0,\pi]$ given by (5.1), (5.5) and (5.6) occurs. If $\rho < \rho_*$ we have coagulation, i.e., the collision results in the formation of a larger particle with mass $M_+ = m_1 + m_2$ and velocity $\vec{V} = (m_1\vec{v}_1 + m_2\vec{v}_2)/M_+$.

This analysis of the pair collision process applies not only to the potential (5.7), but also to other potentials $U(r)$ for which $U(r)r^2 \to -\epsilon$ as $r \to 0$. See [2] for details.

6 Collision integrals for elastic scattering

For a repulsive pair potential defining a scattering cross-section as discussed above, the collision term for the Boltzmann equation is

$$Q(f,f)(v) := \int_{\mathbb{R}^3\times S^2} dw\, dn\, \sigma(|u|,\phi)|u|\,[f(v')f(w') - f(v)f(w)] \tag{6.1}$$

where $u = v - w$, the arguments x and t have been suppressed, and the post-collisional velocities v' and w' are given by

$$v' = \frac{1}{2}(v + w + |u|n), \quad w' = \frac{1}{2}(v + w - |u|n).$$

Here, $n = (\sin\theta\cos\alpha, \sin\theta\sin\alpha, \cos\theta)$, $\quad dn = \sin\theta d\theta d\alpha$, $0 \le \theta \le \pi$, $\quad 0 \le \alpha \le 2\pi$. Compare with (4.1).

In preparation for writing the collision integral for attractive potentials, we rewrite (6.1) in the more compressed form

$$Q(f,f)(v) = \int_{\mathbb{R}^3} dw\,|v - w|\,\Psi\left(\frac{v+w}{2}, v - w\right) \tag{6.2}$$

with

$$\Psi(V,U) = \int_{S^2} dn\, \sigma(|U|,\phi)\,\{F(V,|U|n) - F(V,U)\} \tag{6.3}$$

and $F(V,U) = f(V + U/2)f(V - U/2)$.

Note that the original form of the Boltzmann collision integral includes an integration over the impact parameter $\rho \in (0,\infty)$, and *not* over the scattering angle θ. With the integral over ρ and (5.2), the integral in (6.3) becomes (we omit the argument V)

$$\Psi(U) = \int_0^\infty d\rho\,\rho \int_0^{2\pi} d\alpha\,\{F(|U|n) - F(U)\}, \tag{6.4}$$

where $n \in S^2$ is still defined as before, but the polar scattering angle θ is now a function of ρ as defined by the scattering problem which we discussed earlier. If the intermolecular potential $U(r)$ is positive and monotonically decreasing with r, we have a one-to-one correspondence between $0 \le \theta \le \pi$ and $0 \le \rho < \infty$, and (5.1) defines the relationship between (6.3) and (6.4).

In the sequel we consider the integral (6.3) for attractive potentials when the dependence between ρ and θ is given by the formulas (5.1), (5.5) and (5.6). We assume monotone dependence between ρ and φ.

To simplify our notation, we write the integral (6.4) in the compressed form

$$\Psi(U) = \int_0^\infty d\rho\,\rho\, G[\theta(\rho)], \tag{6.5}$$

with $G[\theta(\rho)] = \int_0^{2\pi} d\alpha \, \{F(|U|n) - F(U)\}$, where $n = (\theta(\rho), \alpha)$ is given in spherical coordinates with the polar axis in direction U. Using (5.1), we define ρ_n, $n = 1, 2, \ldots$, by $\rho_1 = \infty$ and $\varphi(\rho_n) = n\pi, \quad n = 2, 3, \ldots$ If there is a $\varphi_{\max}$ such that $\varphi(\rho) \to \varphi_{\max}$ as $\rho \to 0$, and if we set $\varphi_{\max} = N\pi + \varphi_0$ with $0 \le \varphi_0 \le \pi$, we set $\rho_n \equiv 0$ for $n > N$. Using (5.6), we rewrite (6.4) as

$$\Psi(U) = \sum_{k=1}^{\infty} \left\{ \int_{\rho_{2k}}^{\rho_{2k-1}} d\rho \, \rho G[\varphi(\rho) - (2k-1)\pi] + \int_{\rho_{2k+1}}^{\rho_{2k}} d\rho \, \rho G[(2k+1)\pi - \varphi(\rho)] \right\} \quad (6.6)$$

$\varphi(\rho)$ being defined in (5.1).

Let

$$\begin{aligned} \tilde{\rho}^{(2k)}(\theta) &= \rho[(2k-1)\pi + \theta], \quad k = 1, 2, \ldots \\ \tilde{\rho}^{(2k+1)}(\theta) &= \rho[(2k+1)\pi - \theta], \quad k = 1, 2, \ldots \end{aligned} \quad (6.7)$$

$0 \le \theta \le \pi$. We remark that the inverse function $\rho(\varphi)$ exists in view of our assumption that $\varphi(\rho)$ is strictly monotone.

It is now natural to introduce a set of "partial" differential cross section (see (5.2) for comparison)

$$\sigma_n(\theta) = \frac{\tilde{\rho}^{(n)}(\theta)}{\sin\theta} \left| \frac{d}{d\theta} \tilde{\rho}^{(n)}(\theta) \right|, \quad n = 2, 3, \ldots \quad (6.8)$$

and to set $\rho = \tilde{\rho}^{(2k)}(\theta)$ or $\rho = \tilde{\rho}^{(2k+1)}(\theta)$ in each of the integrals in (6.6). Returning to the integration variable $\theta \in [0, \pi]$, the result is

$$\Psi(U) = \int_0^{\pi} d\theta \, \sin\theta \, F(\theta) \, \hat{\sigma}(|U|, \theta), \quad (6.9)$$

with

$$\hat{\sigma}(|U|, \theta) = \sum_{n=2}^{\infty} \sigma_n(|U|, \theta). \quad (6.10)$$

If $\varphi_{\max} < \infty$, the sum in (6.10) is actually a finite sum. For example, $\sigma_n(|U|, \theta) = 0$ for $n = 3, 4, \ldots$ for the Newtonian potential with $\gamma = 1$, while $\sigma_2(|U|, \theta)$ is merely the classical Rutherford cross-section in this case (the integral (6.5) is then actually divergent, a fact we ignore for the current formal discussion).

The partial differential cross-sections $\sigma_n(\theta)$ in (6.8) allow the following physical interpretation: Each function $\sigma_n(\theta)$ is associated with the relative contribution of those trajectories which have exactly $n-1$ intersections with a real axis directed along u.

To summarize, we obtain for an attractive potential $U(r)$ the usual Boltzmann collision integral (6.1), provided that $|U(r)|r^2 \to 0$ as $r \to 0$. The only difference is that we have to use a generalized differential cross-section $\hat{\sigma}(|U|, \theta)$ as given by (6.10) to replace the usual cross-section $\sigma(|U|, \theta)$.

We now discuss the case $U(r) = -\epsilon/r^2$. Elastic scattering is again described by the Boltzmann collision integral in the form (6.2)-(6.4), provided that we integrate over $d\rho$ in (6.4) *not* from 0 to ∞, but from $\rho = \rho_{\min} = \rho_*$ (as given by (5.8)) to ∞. All the above arguments carry over. Inverting the explicit formula (5.8), we find

$$\rho^2 = \rho_*^2 \left[1 - (\pi/\varphi)^2\right]^{-1}, \quad (6.11)$$

hence

$$\rho\frac{d\rho}{d\varphi} = -\frac{\rho_*^2\pi^2}{\varphi^3}\left[1-(\pi/\varphi)^2\right]^{-2} = -\frac{(\pi\rho_*)^2\varphi}{[\varphi^2-\pi^2]^2}. \tag{6.12}$$

From (6.7), (6.8)

$$\begin{aligned}\sin\theta\ \sigma_{2k}(\theta) &= (\pi\rho_*)^2\frac{(2k-1)\pi+\theta}{\{[(2k-1)\pi+\theta]^2-\pi^2\}^2},\\ \sin\theta\ \sigma_{2k+1}(\theta) &= (\pi\rho_*)^2\frac{(2k+1)\pi-\theta}{\{[(2k+1)\pi-\theta]^2-\pi^2\}^2},\end{aligned} \tag{6.13}$$

$k = 1, 2, \ldots$ The formulas (6.10) and (6.13) (note that the convergence in (6.10) is straightforward) define a generalized elastic cross-section $\hat{\sigma}(|U|, \theta)$ for the potential $U(r) = -\epsilon/r^2$, $\epsilon > 0$.

7 Boltzmann-Smoluchowski collision integrals

We focus on the potential

$$U(r) = -\frac{\epsilon}{r^2}, \quad \epsilon > 0. \tag{7.1}$$

As discussed earlier, a classical solution of the scattering problem for this potential exists only for sufficiently large impact parameters $\rho^2 > \rho_*^2 = 2\epsilon/(\mu u^2)$ and reduced mass $\mu = m_1 m_2/(m_1 + m_2)$. If $\rho < \rho_*$, we have to consider a generalized solution of the scattering problem which incorporates a coagulation process. A consequence is that we have to consider particles with different masses in any case, because particle masses change in the collision (coagulation) process.

1. We replace the distribution function $f(x, v, t)$ of a simple gas by a new distribution function $f(m, x, v, t)$, where $m > 0$ denotes a particle mass. The total number of particles is now given by the formula

$$N = \iint dx\, dv \int_0^\infty dm\, f(m, x, v, t). \tag{7.2}$$

2. We assume that a pair collision with an impact parameter $\rho < \rho_*$ results in coagulation of particles, i.e., in the "reaction"

$$(m_1, v_1) + (m_2, v_2) \longrightarrow (M, V) \tag{7.3}$$

with $M = m_1 + m_2$, $V = (m_1 v_1 + m_2 v_2)/M$. Note that if $\rho < \rho_*$ the postcollisional velocity V does not depend on the impact parameter ρ. The coagulation process is therefore completely determined by a total coagulation cross-section

$$\sigma_c(u; m_1, m_2) = \pi\rho_*^2. \tag{7.4}$$

An elastic differential cross-section can now be calculated in the same way as described in Section 6. We set

$$\rho_*^2 = \frac{2\epsilon(m_1 + m_2)}{m_1 m_2 u^2} \tag{7.5}$$

in (6.13) and calculate a differential cross-section $\hat{\sigma}(u, \theta; m_1, m_2)$ by the formulas (6.10-6.13).

We are now ready to write the Boltzmann-Smoluchowski equation for a distribution density $f(m,x,v,t)$:

$$\frac{\partial f}{\partial t} + v\cdot\nabla_x f = Q_B + I_{Sm} \tag{7.6}$$

where Q_B and I_{Sm} denote the Boltzmann and Smoluchowski collision integrals, respectively. As discussed in section 4, a Vlasov term should also be included; we omit this term in the present section for the sake of simplicity.

First calculate cross-sections $\hat{\sigma}(u,\theta;m_1,m_2)$ and $\sigma_c(u;m_1,m_2)$ as described earlier. Then

$$\begin{aligned} Q_B &= \int_0^\infty dm_1 \int_{\mathbb{R}^3\times S^2} dw\, dn\, |U|\hat{\sigma}(|U|,\theta;m,m_1) \\ &\quad \times \{f(m,v')f(m_1,w') - f(m,v)f(m_1,w)\} \end{aligned} \tag{7.7}$$

with $U=v-w$, $U\cdot n = |U|\cos\theta$, $v' = V + \frac{\mu}{m}U\cdot n$, $w = V - \frac{\mu}{m}U\cdot n$, $V = \frac{mv+m_1w}{m+m_1}$, $\mu = \frac{m_1 m}{(m_1+m)}$. This is the obvious generalization of the usual Boltzmann collision integral for particles with different masses.

The Smoluchowski collision integral is

$$\begin{aligned} I_{Sm} &= \frac{1}{2}\int_0^\infty dm_1 \int_0^\infty dm_2 \int_{\mathbb{R}^3\times\mathbb{R}^3} dv_1\, dv_2\, |U|\sigma_c(|U|;m_1,m_2) \\ &\quad f(m_1,v_1)f(m_2,v_2)\,\delta(m_1+m_2-m)\,\delta\left(\frac{m_1v_1+m_2v_2}{m_1+m_2} - v\right) \\ &\quad -f(m,v)\int_0^\infty dm_1 \int dw\, f(m_1,w)|v-w|\,\sigma_c(|v-w|;m_1,m_2) \end{aligned} \tag{7.8}$$

with $U = v_1 - v_2$. The factor $1/2$ in front of the first integral gives the correct number of inelastic collisions. We have omitted the arguments x and t throughout.

The cross sections $\hat{\sigma}(|U|,\theta;m_1,m_2)$ is a symmetric function of the masses m_1 and m_2. The presented form of the coagulation term is convenient for the calculation of inner products

$$(\psi, I_{Sm}) = \int_0^\infty dm \int_{\mathbb{R}^3} dv\, \psi(m,v) I_{Sm}(m,v)$$

with test functions ψ. After some elementary transformations, one finds

$$\begin{aligned} (\psi, I_{Sm}) &= \frac{1}{2}\int dm_1 \int dm_2 \int_{\mathbb{R}^3\times\mathbb{R}^3} dv_1\, dv_2\, f(m_1,v_1)f(m_2,v_2)|U|\sigma_c(|U|;m_1,m_2) \\ &\quad \times \left[\psi\left(m_1+m_2, \frac{m_1v_1+m_2v_2}{m_1+m_2}\right) - \psi(m_1,v_1) - \psi(m_2,v_2)\right], \end{aligned}$$

$U = v_1 - v_2$, from which conservation of mass and momentum readily follow:

$$(m, I_{Sm}) = 0 = (mv, I_{Sm}).$$

We conclude by presenting explicit formulas for (7.1). Let (see (6.10-6.13))

$$\begin{aligned} g(\theta;m_1,m_2) &= 2\frac{\pi^2\epsilon(m_1+m_2)}{m_1m_2}g(\theta), \\ \sin\theta\, g(\theta) &= \sum_{k=1}^\infty \left\{\frac{(2k-1)\pi+\theta}{\{[(2k-1)\pi+\theta]^2-\pi\}^2} + \frac{(2k+1)\pi-\theta}{\{[(2k+1)\pi-\theta]^2-\pi\}^2}\right\}, \\ p(m_1,m_2) &= 2\frac{\pi\epsilon(m_1+m_2)}{m_1m_2}. \end{aligned} \tag{7.9}$$

The collision terms in (7.6) then become

$$Q_B = \int_0^\infty dm_1 \int_{\mathbb{R}^3\times\mathbb{R}^3} dw\, dn \frac{g(\theta; m_1, m_2)}{|v-w|} \times \{f(m,v')f(m_1,w') - f(m,v)f(m_1,w)\} \qquad (7.10)$$

and

$$I_{Sm} = \frac{1}{2}\int_0^m ds \int_{\mathbb{R}^3} du \frac{p(s,m-s)}{|U|} f\left(s, v + \frac{s}{m}U\right) f\left(m-s, v - \frac{m-s}{m}U\right) - f(m,v)\int_0^\infty ds \int dw \frac{p(m,s)}{|v-w|} f(s,w), \qquad (7.11)$$

where v' and w' are defined after (7.7).

For generalizations to other potentials, the two-dimensional case and additional details the reader is referred to [2].

8 Concluding remarks

We have presented a survey on kinetic equations for particle systems in which the particles interact via pair potentials $\pm\epsilon/r^n$, with $\epsilon > 0$ and $1 < n \leq 2$. The emphasis of the discussion was on the case $-\epsilon/r^2$. i.e., the attractive (stellar-dynamic) potential known as the Manev correction. We showed that for this potential, the kinetic equation depicting the time evolution of a particle cloud should include Vlasov interaction, Boltzmann collision and Smoluchowski coagulation integrals, and we derived the corresponding cross-sections.

If one considers the full Manev potential $-\gamma/r - \epsilon/r^2$, in which the term $-\epsilon/r^2$ may be interpreted as a quasirelativistic correction to the Newtonian potential, it follows that the corresponding kinetic equations should include corrective Boltzmann collision and Smoluchowski coagulation terms. A possible application of such equations would be to the dynamics of globular clusters, where close encounters between stars are sufficiently frequent to justify the inclusion of diffusive collision terms in the kinetic equation (in astrophysical applications, the Fokker-Planck collision term [4] is usually employed in this context).

9 Acknowledgement

This work was supported by the Natural Sciences and Engineering Research Council of Canada under operating grant nr. 7847, and by an internal research grant of the University of Victoria. I am grateful to the Fields Institute and to Anna Lawniczak for organizing a workshop on hydrodynamic limits in the fall of 1998. I am indebted to Florin Diacu for introducing me to the Manev potential in the context of $N-$ body problems.

References

[1] A.V. Bobylev, P. Dukes, R. Illner and H.D. Victory, On Vlasov-Manev Equations, I: Foundations, Properties and Non-Global Existence, Journal Stat. Phys. 88, 885-912 (1997)

[2] A. V. Bobylev and R. Illner, Collision Integrals for Attractive Potentials, Journal Stat. Phys. 95 (3/4), 633-650 (1999)

[3] A. V. Bobylev and G. Vilasi, Projective Invariance for Classical and Quantum Systems, J. Gp. Th. Phys. 3 (1), (1995)

[4] C. Cercignani, Theory and Application of the Boltzmann Equation. Springer-Verlag, New York (1988)

[5] R. Illner, H. D. Victory, P. Dukes and A. V. Bobylev, On Vlasov-Manev Equations, II: Local Existence and Uniqueness, Journal Stat. Phys. 91 (3/4), 625-654 (1998)
[6] L. D. Landau, E. M. Lifshitz, Mechanics. Pergamon Press, Oxford-London-Paris (1960)
[7] G. Manev, La gravitation et le principe de l'égalité de l'action et de la réaction, C.R.A.S. 178, 2159-2161 (1924)
[8] G. Manev, Die Gravitation und das Prinzip von Wirkung und Gegenwirkung, Zeitschrift für Physik 31, 786-802 (1925)
[9] G. Manev, Le principe de la moindre action et la gravitation, C.R.A.S. 190, 963-965 (1930)
[10] G. Manev, La gravitation et l'énergie au zéro, C.R.A.S. 190, 1374-1377 (1930)
[11] I. Newton, in Principia, Book I, Article IX, Theorem IV, Corollary 2
[12] R. Steacy, private communication, manuscript in preparation

Fields Institute Communications
Volume **27**, 2000

Free Boundary Problem and Hydrodynamic Limit

Jeremy Quastel
Department of Mathematics
University of Toronto
Sidney Smith Hall
100 St. George Street
Toronto, ON M5S 3G3 Canada
quastel@math.utoronto.ca

The following problem comes up in nuclear waste management.[1] Radioactive waste is placed in a container and buried. Although the container may be very well designed, there is still some leakage, and so radioactive particles leave the container (very slowly we hope!) and then, as a rough approximation, perform independent Brownian motions. In order to contain them, chemical traps are distributed around the area. When a radioactive particle hits a trap it is destroyed. However, the traps also become saturated after a certain number of particles hit them. To keep matters simple we will assume here a homogeneous distribution of traps. The problem is, given the rate of production of particles and the density of traps, what is the shape and size of the set of saturated traps after a long time?

If you believe the shape is spherical, then the radius $r(t)$ as a function of time can be computed. If $c(t)$ is the rate of creation of particles at time t then the expected number of particles which hit the boundary of the sphere up to time T is

$$F(T) = \int_0^T P\left(\sup_{s\le t\le T} |\beta_t|/r(t) \ge 1\right) c(s)ds \tag{0.1}$$

where β_t is a Brownian motion starting at 0 at time s. On the other hand the traps in a sphere of radius $r(T)$ will have eaten about $\alpha\omega_d r^d(T)$ particles. Here α is the density of traps which should really be counted according to how many particles they can destroy before they are saturated. The constant ω_d is the volume of the sphere of radius one in dimension d. So given $c(t)$ and α, the correct $r(t)$ is the one for which $F(T) = \alpha\omega_d r^d(T)$. It looks a little daunting until you realize that (0.1) can be written as

$$F(T) = \int_0^T c(t)dt - \int \rho(x,t)dx$$

2000 *Mathematics Subject Classification.* Primary 60K35; Secondary 82C41.

[1] I learned about this application from Gerard BenArous

where ρ is the solution of the Stefan problem[2]

$$\begin{aligned} \frac{\partial \rho}{\partial t} &= \Delta\rho + c(t)\delta_0, && \text{in } \{\rho > 0\}, \\ \rho &= 0 && \text{on } s(x) \geq t, \\ \nabla\rho \cdot \nabla s &= -\alpha && \text{on } s(x) = t. \end{aligned} \tag{0.2}$$

If $c(t) = t^{(d-2)/2}$ then in any dimension there is an explicit solution and the desired radius turns out to be $K\sqrt{t}$ where $\exp\{-K^2/4\} = \alpha K^d \pi^{d/2}/\Gamma(d/2)$. If $c(t) \equiv 1$ then $r(t) \approx t^{1/d}$ if $d \geq 3$ and $r(t) \approx \sqrt{t \log t}$ if $d = 1$. In $d = 2$ it is just as in the previous line.

On the other hand we have no particular reason to believe that the shape is a sphere, or for that matter that there is any sort of shape at all. So that is the real problem.

We could put the system on a lattice. In other words we replace the physical space $\mathbf{R}^d$ by $\mathbf{Z}^d$ and for each site x in the lattice let ζ_x and ξ_x denote the number of particles and the number of traps in a box of side length one around x. Then we approximate the original model by having the particles produced at the origin at rate $c(s)$ and then perform independent continuous time symmetric random walks on the lattice until they reach a site x where $\xi_x \geq 1$. Then that particle is destroyed and ξ_x is reduced to $\xi_x - 1$.

Such models have been studied by Lawler, Bramson and Griffeath [1] under the name Internal Diffusion Limited Aggregation. In particular, they verified the spherical shape in the case $c(s) \equiv 1$ and $\xi_x \equiv 1$ in dimensions $d \geq 3$. This was extended to $d \leq 2$ by Gravner in joint work with the author [2] and is presently being extended to the case where ξ_x is a stationary random field by Ramirez and the author. From now on we will stick to the case $\xi_x \equiv 1$ in order to illustrate the main ideas. First we want to explain the dimensional dependence of the results. In fact the dimension is not the key point, but the asymptotic behaviour of

$$t^{-d/2} \int_0^t c(s)ds.$$

If this goes to zero as t gets large, then we say the model is subcritical. If it goes to infinity, we say the model is supercritical. In between is the critical case. The density inside the occupied set, in either the Stefan problem, or the Internal DLA, is asymptotically zero in the subcritical case, infinite in the supercritical case, and non-trivial in the critical case. The method of Lawler, Bramson and Griffeath, which is by comparison with a discrete time version of the model, works only in the subcritical regime. The method of Gravner and the author is by comparison with the Stefan problem. Although this comparison is only correct in the critical regime, the Lawler-Bramson-Griffeath method can be turned around and therefore it can be extended to the subcritical regime and recovers even the shape theorems for the discrete time versions. Not much is known in the supercritical regime except in one dimension where there is not much of a shape theorem to prove.

1 Critical case

Let us consider the critical case of Internal DLA. Particles are produced at the origin of $\mathbf{Z}^d$ at rate $c(t) = t^{(d-2)/2}$ and then perform independent random walks until they reach a trap at which point the two are mutually destroyed. The number

[2] We have normalized things so that our Brownian motions have generator Δ.

of particles at x at time t is called $\zeta_x(t)$ and the number of traps at x at time t is called $\xi_x(t)$. Initially $\xi_x(0) = 1$ for each site. What we want to prove is a law of large numbers for the diffusively rescaled density of particles and set of saturated traps: In an appropriate weak sense

$$\zeta_{\lfloor\epsilon^{-1}x\rfloor}(\epsilon^{-2}t) \to \rho(x,t)$$

$$A_t^\epsilon = \{x \in \mathbf{R}^d \ : \ \xi_{\lfloor\epsilon^{-1}x\rfloor}(\epsilon^{-2}t) = 0\} \to \mathcal{A}_t = \{x \in \mathbf{R}^d \ : \ s(x) \le t\}$$

as $\epsilon \to 0$ where $\rho(x,t)$, $s(x)$ is the solution of Stefan problem (0.2).

There are now several general methods to prove such hydrodynamic scaling limits. The most popular are based on the evolution of the relative entropy of the system with respect to global or local equilibrium measures. Ignoring for a moment the creation of particles at the origin, our model has two mutually singular equilibrium classes. The first has a trap at every site and no particle in sight, and the second class of invariant measures has no trap anywhere, while the particles are distributed according to a product of Poisson distributions with a fixed density ρ which parametrizes the class. Unfortunately the entropy methods are not well adapted to systems such as ours, where the key object of study is an unknown boundary separating regions in which the system is locally in one of two such equilibrium classes which have infinite entropy with respect to each other.

The proof is based on the following clever transformation of Stefan problem (0.2) which is well known to researchers in that field. If we define

$$u = \rho + \mathbf{1}(\rho > 0) \qquad \textit{and} \qquad \lambda(u) = (u-1) \vee 0,$$

so that $\rho = \lambda(u)$ then if $\rho(x,t)$, $s(x)$ is a solution of (0.2) an application of Green's theorem shows that $u(x,t)$ is a weak solution of

$$\frac{\partial u}{\partial t} = \Delta\lambda(u) + c(t)\delta_0. \tag{1.1}$$

In fact (1.1) only has one solution, and the number of unknown functions has magically been reduced from two to one, so it is definitely an improvement over (0.2). To see that there is only one solution, we notice that it is in the form of a degenerate porous medium equation and the porous medium equation has the nice property that it contracts the H_{-1} norm,

$$\|f\|_{-1}^2 = \int f(-\Delta^{-1})f dx.$$

In fact if u and u' are two solutions to (1.1) then one can easily check that

$$\|u(T) - u'(T)\|_{-1}^2 = -\int_0^T \int (u-u')(\lambda(u) - \lambda(u'))dxdt. \tag{1.2}$$

The integral on the right is non-negative whenever λ is a non-decreasing function of u.

We can try a similar trick for the Internal DLA. Let

$$\eta_x = \zeta_x + (1 - \xi_x)$$

be the number of live or dead particles at site x in the lattice. Then η by itself is a Markov process and its generator is

$$Lf(\eta) = \sum_{x,|e|=1} \lambda(\eta_x)\left(f(\eta^{x,x+e}) - f(\eta)\right) + t^{(d-2)/2}\left(f(\eta_+^0) - f(\eta)\right).$$

Here $\eta^{x,y}$ means the new configuration after one particle has jumped from x to y, i.e., $\eta^{x,y}_x = \eta_x - 1$, $\eta^{x,y}_y = \eta_y + 1$ and $\eta^{x,y}_z = \eta_z$ otherwise. Similarly η^0_+ means the new configuration after one new particle has been created at the origin. Notice that the process takes the form of a degenerate zero-range model.

The idea, which first appeared in work of Chang and Yau [3] in a different context, is to compare the microscopic system to the macroscopic one using the H_{-1} norm. We work on the lattice $\epsilon\mathbf{Z}^d$ on which we have the Internal DLA model η^ϵ running at rate ϵ^{-2}. We compare it to the lattice version of the Stefan problem (1.1),

$$\frac{\partial u^\epsilon}{\partial t} = \Delta_\epsilon \lambda(u^\epsilon) + dP^\epsilon, \tag{1.3}$$

where the Poisson process $P^\epsilon(t)$ is just the number of particles created in the Internal DLA model up to time t, and the lattice Laplacian is given by

$$\Delta_\epsilon f_x = \epsilon^{-2} \sum_{|e|=\epsilon} (f_{x+e} - f_x).$$

One can show that as $\epsilon \to 0$ the solution u^ϵ converges to the solution u of (1.2), the Poisson process being set up to approximate the creation term $c(t)\delta_0$. We also define the H_{-1} norm of a function on $\epsilon\mathbf{Z}^d$ by

$$\|f\|^2_{-1,\epsilon} = \epsilon^d \sum_{x\in\epsilon\mathbf{Z}^d} f_x(-\Delta_\epsilon^{-1} f)_x.$$

Now you apply the generator to the H_{-1} of the difference of u^ϵ and η^ϵ and after some computation you obtain the following remarkable formula,

$$E[\|\eta^\epsilon(T) - u^\epsilon(T)\|^2_{-1,\epsilon}] = -2\int_0^T \epsilon^d \sum_x E[V(\eta^\epsilon_x(t), u^\epsilon_x(t))]dt \tag{1.4}$$

where

$$V(n,u) = (n-u)(\lambda(n) - \lambda(u)) - \lambda(n).$$

It looks as if something has gone wrong. On the right hand side the integrand is the same sort of non-negative object as in (1.2) except that now it has an extra term $-\lambda(n)$ appended at the end which seems to spoil everything. Of course this term is really very important. If it was not there then we would have shown that for fixed ϵ the H_{-1} norm of the difference of η^ϵ and u^ϵ is zero which is patently false because we expect that inside the occupied set the distribution of the particles is more or less a product of Poissons whose parameter is approximately the solution of the Stefan problem. Therefore the convergence can only be weak and correct in the limit as ϵ is sent to zero.

What happens is that because of the time integral, which is really over an enormous time ϵ^{-2} and the fact that typical sites are far from the origin we can pretend that the expectation of V is being taken locally with respect to an invariant measure of the process without creation. Furthermore we can transfer whatever smoothness can be obtained for u^ϵ into some spatial averaging, so the invariant measure can also be assumed to be translation invariant. One can adapt known coupling arguments [4] to show that all such invariant measures fit into the two classes described a few paragraphs ago and the punch line is that for any measure μ in those classes and any u whatsoever,

$$E_\mu[V(\eta_0, u)] \le 0.$$

The reason it happens is that the extra term $-\lambda(n)$ is exactly the right thing to compensate for the fact that if N is a Poisson random variable with expectation m then $E[N^2] = m^2 + m$. The main result is

Theorem *Let $A_t \subset \mathbf{Z}^d$ be the set of saturated obstacles and $\zeta_x(t)$ the particle count in the internal DLA process with initially one obstacle at each site, with creation of particles at the origin at rate $c(t) = t^{(d-2)/2}$ and particles jumping at rate one to each nearest neighbour site until they reach an obstacle, where the two are mutually annihilated. Then $\epsilon A_{\epsilon^{-2}t}$ conveges to a ball of radius $K\sqrt{t}$ where K is the solution of (0.2) in the sense that $\epsilon^d \times$ the number of sites in the symmetric difference of $\epsilon A_{\epsilon^{-2}t}$ and $\{x \in \epsilon\mathbf{Z}^d : |x| \le K\sqrt{t}\} \to 0$ as $\epsilon \to 0$. Furthermore $\zeta_{\lfloor \epsilon^{-1}x \rfloor}(\epsilon^{-2}t)$ converges weakly as $\epsilon \to 0$ to the solution $\rho(x,t)$ of Stefan problem (0.2).*

The case $d = 2$ settles a conjecture of Lawler, Bramson and Griffeath. The details of the proof can be found in [2]. We also learned at the meeting that Funaki [6] has recently studied the same model without creation and shown convergence to the Stefan problem. We note that our method can be adapted to allow the particles to interact to some extent before they are killed. One replaces the function $\lambda(n)$ in the generator L by $g(n-1) \vee 0$ where g is some general increasing function with linear growth. Then $\lambda(u)$ in the Stefan problem (1.1) is replaced by $\gamma(u-1) \vee 0$ where γ is an appropriate function depending on the choice of g and the result is a non-linear version of the Stefan problem. One can also have several creation sites and there are analogous results in that case as well.

2 Supercritical case

In the supercritical case we do not know how to prove any shape theorem so we are restricted to $d = 1$. We let $c(t) = t^\alpha$ with $\alpha > -1/2$ and compute the size of the occupied set by studying the asymptotics of (0.2). Note first of all that

$$u(x,t) = \int_0^t |t-s|^\alpha \, \frac{e^{-x^2/4s}}{\sqrt{4\pi s}} \, ds$$

satisfies the equation $u_t = u_{xx} + \delta_0$. We look for a function $r(t)$ such that $\hat\rho(x,t) = u(x,t) - u(r(t),t)$ for $|x| \le r(t)$ is an approximate solution of (0.2). We need $-u_x(r(t),t) \sim r'(t)$ and $\partial_t u(r(t),t) \sim 0$. Then we can construct out of $\hat\rho$ appropriate sub and super-solutions to show that $r(t)$ is the real size of the boundary. Differentiating and changing variables we get

$$-u_x(x,t) = \frac{t^\alpha}{\sqrt{\pi}} \int_{x^2/4t}^\infty \left| u - \frac{x^2}{4t} \right|^\alpha u^{-(\alpha+1/2)} e^{-u} du.$$

Using the fact that $\int_a^\infty |u-a|^\alpha u^{-(\alpha+1/2)} e^{-u} du \sim a^{-(\alpha+1/2)} e^{-a}$ as $a \to \infty$ one can check that

$$r(t) = \sqrt{4t \log \left\{ t^{-d/2} \int_0^t c(s) ds \right\}}$$

is the radius of the occupied set. The case $c(t) = 1$ was previously obtained [2], [5] by much more complicated methods.

3 Subcritical case

The discrete time version of Internal DLA is a Markov chain $A(0), A(1), \ldots$ whose state space is the set of connected sets in the lattice $\mathbf{Z}^d$ containing 0 and whose transition probabilities are given by $p(A,B) = 0$ unless B can be obtained from A by adding on one site x in its boundary, in which case $p(A,B)$ is the probability that a random walk starting at the origin hits the boundary of A at x.

Let A_t denote the occupied set (set of saturated obstacles) in the continuous time version of Internal DLA with some creation rate $c(t)$. We now give a rough sketch of the argument of Lawler, Bramson and Griffeath [1] comparing A_t with the discrete time version $A(n)$. One numbers the particles according to their time of birth, and lets $A_t(n)$ be the occupied set of the first n particles, with the rule that if an older particle arrives at a site where an younger particle has died, the older particle replaces the younger one as dead at that site and the younger particle goes off on its merry way. $A_t(n)$ is always a subset of both A_t and $A(n)$ and, for fixed n, $A_t(n)$ increases to $A(n)$ as t increases to infinity since each of the first n particles will after some time have found its resting place. If we let $P(t)$ be the number of particles created up to time t in the continuous model, then $P(t)$ is about

$$n(t) \doteq \int_0^t c(s)ds$$

and $A_t(P(t)) = A_t$ so we have

$$A_t(n(t)) \simeq A_t$$

the only error coming from the surplus or surfeit of particles produced up to time t. By $\simeq$ we mean that the measure of the symmetric difference normalized by the measure of the union, goes to zero as $t \to \infty$. In order to compare A_t with $A(n)$ the key point is therefore whether

$$A_t(n(t)) \simeq A(n(t))$$

which is a way of saying that the density of live particles is going to zero, or, in other words, that we are in the subcritical case. The precise argument can be found in reference 6 and the conclusion is that the shape theorems for $A(n)$ and A_t can be deduced from each other. Lawler, Bramson, and Griffeath then used probabilistic potential theory methods to show that $A(n)$ is asymptotically a sphere. Actually, their argument works in the much stronger Hausdorff metric, i.e. they show that for sufficiently large n, $n^{-1/d}A(n)$ contains any sphere centered at the origin of volume less than one, and is contained in any sphere centered at the origin of volume greater than one.

What we want to do here is deduce the shape theorems for both $A(n)$ and A_t from a hydrodynamic limit, complementing the Lawler-Bramson-Griffeath argument.

We introduce a 'soft' version of Internal DLA with creation rate $c(t)$. Particles are produced at the origin in $\mathbf{Z}^d$ at rate $\epsilon^{-d}c(t)$ and perform independent random walks at rate ϵ^{-2} until they reach a trap, at which point the two are mutually annihilated as usual. We are going to compare the soft model to the original model which has creation at rate $c(t)$ and walks at rate 1. The soft model is a critical, renormalized version of the original model and, in fact, the hydrodynamic methods apply to it without modification showing that the density and occupied set converge to those of the subcritical Stefan problem (0.2): If we let A_t^ϵ denote the occupied set of the soft model and $\mathcal{A}_t$ denote the occupied set of Stefan problem (0.2) then what we have is that as ϵ goes to zero,

$$\epsilon A_t^\epsilon \simeq \mathcal{A}_t.$$

We know that $\mathcal{A}_t$ is spherical, but we don't know its radius explicitly. However, from the subcriticality it is clear that as t goes to infinity, $n(t)^{-1/d}\mathcal{A}_t$ converges to a ball of volume one. Now let $P^\epsilon(t)$ be the number of particles created in the soft model up to time t. By the law of large numbers, $P^\epsilon(t)$ is approximately $n(\epsilon,t) \doteq \epsilon^{-d}n(t)$ and we have

$$A_t^\epsilon(n(\epsilon,t)) \simeq A_t^\epsilon$$

and therefore

$$n(\epsilon,t)^{-1/d}A_t^\epsilon(n(\epsilon,t)) \simeq \text{A } \textit{ball of volume one}$$

as ϵ goes to zero then t goes to infinity. But also from the subcriticality we have that

$$A_t^\epsilon(n(\epsilon,t)) \simeq A(n(\epsilon,t))$$

from which we can conclude that as $n \to \infty$,

$$n^{-1/d}A(n) \simeq \text{A } \textit{ball of volume one.}$$

The reason that all subcritical models have the same asymptotics is that if the density of live particles is going to zero the growth of the boundary is essentially by harmonic measure from the origin. In the critical case where there is a non-trivial density of particles in the region, the growth of the boundary is essentially by the average of the harmonic measure from all points inside the region, weighted by the density, which is itself a dynamic object.

4 Future directions

The most important open problem about internal DLA models is the size of the boundary fluctuations. The only result is Lawler's [7] which says that for $d \geq 2$ the fluctuations of $A(n)$ are most of order $n^{1/3}$ (with logarithmic corrections). It is believed that the fluctuations should be logarithmic in $d = 2$ and of order 1 in $d \geq 3$.

An open problem we have already mentioned is the shape theory in the supercritical regime in $d \geq 2$. If the particles are injected sufficiently rapidly ($c(t)$ grows sufficiently rapidly), one expects that the shape will no longer be spherical but will be a rounded off diamond.

In the case where the number of obstacles ξ_x at the initial time form a random field, the stationary case with $E[\xi_x] < \infty$ can be analysed with our methods. An interesting open problem is what happens in the 'deep well' case $E[\xi_x] = \infty$. We know of no results in this direction.

References

[1] Lawler, G., Bramson, M., Griffeath, D., *Internal diffusion limited aggregation*, Ann. Prob. **20** (1992), no. 4, 2117–2140.

[2] Gravner, J., Quastel, J., *Internal DLA and the Stefan problem*, preprint.

[3] Chang, C.C., Yau, H.-T., *Fluctuations of one-dimensional Ginzburg-Landau models in nonequilibrium*. Comm. Math.. Phys. **145** (1992), no. 2, 209–234., Yau, H.-T., *Metastability of Ginzburg-Landau model with a conservation law*. J. Statist. Phys. **74** (1994), no. 3-4, 705–742.

[4] Liggett, T., *Coupling the simple exclusion process*, Ann. Prob., **4** (1976), 339–356., Andjel E., *Invariant measures for the zero-range processes*, Ann. Prob. **10** 525–547 (1982).

[5] BenArous, G., Ramirez, A., *Diffusion and saturation processes in random media*, preprint.

[6] Funaki, T., preprint.

[7] Lawler, G., *Subdiffusive fluctuations for internal diffusion limited aggregation*, Ann. Prob., **23** (1995), 71-86.

Fields Institute Communications
Volume **27**, 2000

A Variational Coupling for a Totally Asymmetric Exclusion Process with Long Jumps but No Passing

Timo Seppäläinen
Department of Mathematics
Iowa State University
Ames, Iowa 50011, USA
seppalai@iastate.edu

Abstract. We prove a weak law of large numbers for a tagged particle in a totally asymmetric exclusion process on the one-dimensional lattice. The particles are allowed to take long jumps but not pass each other. The object of the paper is to illustrate a special technique for proving such theorems. The method uses a coupling that mimics the Hopf-Lax formula from the theory of viscosity solutions of Hamilton-Jacobi equations.

1 Introduction

The purpose of this paper is to explain a technique for deriving scaling limits that applies to certain totally asymmetric particle processes. The example we use to illustrate the method is a one-dimensional exclusion process whose particles are allowed to take arbitrarily long jumps, as long as particles do not pass each other. Thus the ordering of particles is preserved, and the process models traffic on a single-lane highway. To the author's knowledge, the hydrodynamics of this particular process have not been addressed before.

The key point of the technique is this: There exists a coupling that expresses a process with general initial conditions in terms of a family of processes that have a simple structure. The evolution from a general initial configuration is constructed from the simple cases by a variational formula, hence the term "variational coupling". The benefit of the scheme is that the limit behavior of the simple processes is tractable, and through the coupling one obtains results for the general process.

This paper is intended as an introduction to the technique. Fairly complete proofs are provided, and we have not strived to obtain the best possible result for the process under study. Some extensions of the theorem of the paper and shortcomings of the method are addressed in Section 6. A review of limit theorems for tagged particles and references to literature can be found in [3].

2000 *Mathematics Subject Classification.* Primary 60K35; Secondary 82C22.
Research supported in part by National Science Foundation grant DMS-980108.

Notational remarks. For a real number x, $[x]$ denotes the largest integer less than or equal x. $\mathbf{Z}_-$ is the set of nonpositive integers $\{\ldots,-3,-2,-1,0\}$.

2 Description and construction of the process

The process consists of particles that move on the one-dimensional lattice $\mathbf{Z}$. At most one particle is permitted on each site. (This is the meaning of the name "exclusion process".) We label the particles by integers $i \in \mathbf{Z}$. The state of the process is a sequence of integers $\sigma = (\sigma(i) : i \in \mathbf{Z})$, where $\sigma(i)$ is the location of particle i. We shall also use "$\sigma(i)$" as the name of the particle, because typically we need to discuss simultaneously several different processes denoted by different Greek letters. The dynamics will be such that the ordering of particles is preserved. So we can assume that

$$\sigma(i) + 1 \le \sigma(i+1) \text{ for all } i \in \mathbf{Z}. \tag{2.1}$$

This condition contains both the exclusion rule and the no-passing rule.

It is also convenient to allow the values $\pm\infty$ for $\sigma(i)$. Assumption (2.1) remains valid. If $\sigma(i) < \infty = \sigma(i+1)$, then $\sigma(i)$ is the rightmost particle on the lattice $\mathbf{Z}$ because $\sigma(j) = \infty$ for all $j > i$. Similarly there is a leftmost particle if some $\sigma(j) = -\infty$.

To summarize, the state space of the process is the set of $\mathbf{Z} \cup \{\pm\infty\}$-valued sequences σ that satisfy (2.1).

For the dynamics, assume given a sequence β_k, $k \ge 1$, of nonnegative real numbers. These are the rates. For each k and each i, particle $\sigma(i)$ independently attempts to jump k steps to the right at exponential rate β_k. If there are k empty lattice spaces in front of $\sigma(i)$, this jump can be executed. If there are not, in other words, if $\sigma(i+1) \le \sigma(i) + k$, then $\sigma(i)$ jumps as much as it can, to the location $\sigma(i+1) - 1$ right behind the next particle. The jumps occur independently for all particles $\sigma(i)$ and all jump sizes k.

For nonnegative integers m, define

$$B_m = \sum_{k=1}^{\infty} k^m \beta_k \,. \tag{2.2}$$

We assume

$$B_2 < \infty. \tag{2.3}$$

Then also $B_0 < \infty$, which guarantees that there are finitely many jump attempts in finite time intervals almost surely. $B_2 < \infty$ implies that the distance traveled by any particle during a finite time interval has a finite second moment.

To rigorously construct the process, let $\{\mathcal{D}_i : i \in \mathbf{Z}\}$ be an i.i.d. collection of homogeneous rate B_0 Poisson point processes on $[0,\infty)$. In other words, a time interval $[t, t+h)$ has a Poisson$(B_0 h)$-distributed random number of points (or *epochs*) from the process $\mathcal{D}_i$. For each process $\mathcal{D}_i$, label the points independently by integers $k \ge 1$, with probability β_k / B_0 for label k. Let $\mathcal{D}_i^k$ denote the point process of k-epochs (that is, the points with label k) from process $\mathcal{D}_i$. It follows that $\{\mathcal{D}_i^k : i \in \mathbf{Z},\, k \ge 1\}$ is a collection of independent homogeneous Poisson point processes on $[0,\infty)$, and the rate of $\mathcal{D}_i^k$ is β_k. If $\beta_k = 0$ the point process $\mathcal{D}_i^k$ is empty for all i.

The idea of the construction is that particle $\sigma(i)$ reads its jump commands from the Poisson process $\mathcal{D}_i$, and attempts to execute a k-step jump at each epoch of $\mathcal{D}_i^k$.

Let $\sigma(i,t)$ denote the location of particle $\sigma(i)$ at time $t \geq 0$, and $\sigma(t) = (\sigma(i,t) : i \in \mathbf{Z})$ the entire state at time t. Assume that a deterministic initial configuration $\sigma(0) = (\sigma(i,0) : i \in \mathbf{Z})$ has been specified. Informally, the rule of evolution is this:

Jump rule. Suppose τ is an epoch of $\mathcal{D}_i^k$. Assume that the state $\sigma(t)$ has been defined for $t < \tau$. At time τ set

$$\sigma(i,\tau) = \min\{\sigma(i,\tau-) + k\,,\, \sigma(i+1,\tau-) - 1\}\,. \tag{2.4}$$

After τ, $\sigma(i)$ stays constant until the next epoch in $\mathcal{D}_i$. The same scheme happens simultaneously and independently for all particles $\sigma(i)$.

Since there are infinitely many particles, infinitely many jumps are attempted almost surely in any time interval $(0,\varepsilon)$. In particular, there cannot be a first jump. Consequently some argument is needed to justify the induction in the jump rule, so that formula (2.4) can be used to define the successive locations of particle $\sigma(i)$.

The realizations $\{\mathcal{D}_i^k\}$ for which we can construct the dynamics satisfy these assumptions:

(a) Each $\mathcal{D}_i$ has only finitely many epochs in each bounded time interval.

(b) There are no simultaneous jump attempts.

(c) There are arbitrarily large i_0 and t_0 such that $\mathcal{D}_{-i_0}$ and $\mathcal{D}_{i_0}$ have no epochs in $[0,t_0]$. (2.5)

It is a standard fact of Poisson processes that these assumptions are satisfied by almost every realization $\{\mathcal{D}_i^k\}$.

Assumptions (2.5) allow us to prove that the jump rule defines the evolution unambiguously for all particles $\sigma(i)$ and all times $0 \leq t < \infty$: Given i and t, part (c) of (2.5) gives i_0 and t_0 so that $-i_0 < i < i_0$ and $t_0 > t$, and so that $\mathcal{D}_{-i_0}$ and $\mathcal{D}_{i_0}$ have no epochs in $[0,t_0]$. Consequently $\sigma(-i_0,t) = \sigma(-i_0,0)$ and $\sigma(i_0,t) = \sigma(i_0,0)$ for all $t \in [0,t_0]$, and the evolution of particles $\sigma(j)$ for $-i_0 < j < i_0$ is isolated from the rest of the process up to time t_0. By parts (a) and (b) of (2.5), the locations $\sigma(j,t)$ for $-i_0 < j < i_0$ and $t \in [0,t_0]$ can be computed by applying the jump rule to the finitely many potential jump times in $\cup_{-i_0<j<i_0}\mathcal{D}_j \cap [0,t_0]$, in their temporal order. In particular, the motion of particle $\sigma(i)$ is defined up to time t.

This way the evolution $\sigma(\cdot) = \{\sigma(t) : t \geq 0\}$ is defined as a function of an arbitrary initial configuration $\sigma(0)$, and of almost every realization of the jump times $\{\mathcal{D}_i^k\}$. If $(\Omega, \mathcal{F}, P)$ is a probability space on which are defined the point processes $\{\mathcal{D}_i^{l_0}\}$ and a random initial configuration $\sigma(0)$, then $\sigma(\cdot)$ is a stochastic process defined on this same probability space. Note: The Poisson point processes have to be independent of the initial configuration $\sigma(0)$.

Constructing particle systems from Poisson processes of potential jump times is a standard procedure, and can be found in general references on particle systems such as [1, 4, 5]. A useful property of this construction is that it preserves orderings between two processes:

Lemma 2.1 *Suppose the probability space $(\Omega, \mathcal{F}, P)$ contains two initial configurations $\sigma'(\cdot,0)$ and $\sigma''(\cdot,0)$ that satisfy $\sigma'(i,0) \leq \sigma''(i,0)$ for all $i \in \mathbf{Z}$ with probability 1. If the processes $\sigma'(\cdot,t)$ and $\sigma''(\cdot,t)$ obey the same point processes $\{\mathcal{D}_i^k\}$, we have $\sigma'(i,t) \leq \sigma''(i,t)$ for all $i \in \mathbf{Z}$ and $t \geq 0$, with probability 1.*

Proof Choose i_0 and t_0 to satisfy part (c) of assumption (2.5). We shall prove that the ordering $\sigma'(i,t) \le \sigma''(i,t)$ holds for $-i_0 \le i \le i_0$ and $t \le t_0$. It holds for $i = \pm i_0$ because particles $\sigma'(\pm i_0)$ and $\sigma''(\pm i_0)$ do not attempt to jump during the time interval $[0, t_0]$. By parts (a) and (b) of (2.5), we may do induction over the jump times. Suppose $\tau \in \mathcal{D}_i^k$ is the first epoch at which the ordering is violated, for $-i_0 \le i \le i_0$ and $\tau \le t_0$. Then we must have $\sigma'(i,\tau) > \sigma''(i,\tau)$ but $\sigma'(i,\tau-) \le \sigma''(i,\tau-)$. Since $\sigma''(i)$ was not allowed to jump as far as $\sigma'(i)$ even though they both attempted a k-jump, it must be that $\sigma'(i+1,\tau-) > \sigma''(i+1,\tau-)$. Since τ is the first time the ordering is violated for indices $-i_0 \le i \le i_0$, it must be that $i+1 > i_0$. Since $i_0 \ge i$, we conclude that $i = i_0$. But $\mathcal{D}_{i_0}$ was chosen to have no epochs up to time t_0. We have reached a contradiction, so there can be no first violation of the ordering for $-i_0 \le i \le i_0$ and $t \le t_0$. Since i_0 and t_0 can be taken arbitrarily large, the proof is complete. □

3 The heuristic behind the method

We are interested in a macroscopic view of the process. By this we mean its large scale behavior, distinct from the local interactions of the individual particles $\sigma(i)$ that represent the microscopic dynamics. The ratio between macroscopic and microscopic scales is given by a parameter n, so that macroscopic space and time units represent n microscopic space and time units. We hope that a deterministic macroscopic description of the dynamics emerges in the limit $n \to \infty$, where the macroscopic and microscopic scales become infinitely separated.

What form should the deterministic macroscopic description take? Suppose we have a function $u(x,t)$ defined for macroscopic space-time variables (x,t) such that the approximate equality

$$u(x,t) \approx n^{-1}\sigma([nx], nt) \tag{3.1}$$

holds with high probability for large n. Since the rate of advance of particle $\sigma([nx], nt)$ depends only on the distance to the next particle on the right, we might expect this same structure also at the macroscopic level, so that

$$u_t = f(u_x) \tag{3.2}$$

for some function f. Equation (3.2) is a first-order partial differential equation of the Hamilton-Jacobi type. It is usually written as $u_t - f(u_x) = 0$, and then the function $-f$ is known as the *Hamiltonian.*

The exclusion rule and (3.1) force $u_x \ge 1$, so f is a function defined on $[1, \infty)$. If the distance to the next particle ahead is one (no empty sites) a particle cannot jump, and hence $f(1) = 0$. The rate of advancement cannot decrease as the distance to the next particle increases, so f ought to be nondecreasing. No particle can advance faster than at rate B_1, so $f \le B_1$. Since f is nondecreasing and bounded above, it appears reasonable to assume that f is concave.

Thus, assuming our heuristic reasoning is reliable, if there is to be a law of large numbers

$$n^{-1}\sigma([nx], nt) \to u(x,t)\,, \tag{3.3}$$

then u should satisfy equation (3.2) with a convex Hamiltonian $-f$. There is a well-known formula for solving (3.2) when the Hamiltonian is convex: Let

$$g(x) = \sup_{v \ge 1}\{vx + f(v)\} \tag{3.4}$$

be the convex conjugate of $-f$. Suppose the initial data for (3.2) is given by a function u_0 on $\mathbf{R}$. Then the *Hopf-Lax formula*

$$u(x,t) = \inf_{y \geq x} \left\{ u_0(y) + tg\left(\frac{x-y}{t}\right) \right\} \tag{3.5}$$

defines the unique viscosity solution of (3.2) with initial data $u(x,0) = u_0(x)$ (Theorem 3, Section 10.3 in [2]). Note: Since $f \geq 0$, (3.4) shows that $g(x) = \infty$ for $x > 0$. That is why we may restrict the infimum in (3.5) to $y \geq x$.

However, our task is not to solve the differential equation (3.2) with a given f. In some sense we already have the solutions, at least in their microsopic form $\sigma([nx], nt)$ because we already constructed the particle process. Instead, our problem is to prove that the process σ admits a macrosopic description of the type (3.2), and to find f and g, if possible. The Hopf-Lax formula suggests a line of attack on our problem.

Suppose we are allowed to feed a single input $u_0 = \varphi_0$ to the system described by (3.5), and measure the output $\varphi(x,t) = u(x,t)$. Can we obtain g? Take

$$\varphi_0(y) = \begin{cases} \infty & \text{if } y > 0, \\ y & \text{if } y \leq 0. \end{cases} \tag{3.6}$$

With $u_0 = \varphi_0$, differentiating inside the braces in (3.5) shows that $\varphi(x,t) = u(x,t) = tg(x/t)$. (The duality (3.4) implies that the slope of g is everywhere at least 1, since this is the lower bound for v.) Once we have g, we can compute the solutions to (3.2) by the Hopf-Lax formula (3.5). In other words, the single evolution $\varphi(x,t)$ started with (3.6) contains the information for constructing *all* evolutions from their initial data.

To understand this point better, let u_0 be an arbitrary initial function for the evolution, and consider this family of special initial data, indexed by $z \in \mathbf{R}$:

$$\varphi_0^z(y) = u_0(z) + \varphi_0(y), \quad y \in \mathbf{R}. \tag{3.7}$$

These are simply translates of the initial data (3.6). Find the evolution from (3.5):

$$\begin{aligned} \varphi^z(x,t) &= \inf_{y \geq x} \left\{ \varphi_0^z(y) + tg\left(\frac{x-y}{t}\right) \right\} \\ &= u_0(z) + \inf_{y \geq x} \left\{ \varphi_0(y) + tg\left(\frac{x-y}{t}\right) \right\} \\ &= u_0(z) + tg(x/t). \end{aligned}$$

We substitute the evolution $\varphi^z(x,t)$ into (3.5) to obtain

$$u(x,t) = \inf_{y \geq x} \varphi^y(x-y,t). \tag{3.8}$$

Equations (3.7) and (3.8) contain the message: A general evolution $u(x,t)$ starting from initial data u_0 can be computed from the family of solutions $\{\varphi^y : y \in \mathbf{R}\}$, whose initial data are translates $\varphi_0^y = u_0(y) + \varphi_0$ of the function φ_0.

Let us move the discussion to the microscopic particle level. We can realize the initial data (3.6) microscopically, in the sense of the limit (3.3), by a natural particle configuration:

$$\sigma(i,0) = \begin{cases} \infty & \text{if } i > 0, \\ i & \text{if } i \leq 0. \end{cases} \tag{3.9}$$

This says that the lattice is empty to the right of the origin, while to the left all sites are occupied, the origin itself included. Since (3.9) is a microscopic version of (3.6), we would expect that the process $\sigma(t)$ with initial configuration (3.9) is a microscopic version of the solution $\varphi(x,t) = tg(x/t)$. Translates of these solutions suffice to express the most general solution, as observed in (3.8).

The key question becomes: Is there a microscopic principle that corresponds to (3.8) and expresses a process with a general initial configuration in terms of a family of processes of the type that start from (3.9)? The construction we seek is the coupling we now turn to.

4 The coupling

Assume given an arbitrary deterministic initial particle configuration $\sigma(0) = (\sigma(j,0) : j \in \mathbf{Z})$, and let $\sigma(j,t)$ denote the process constructed in terms of the Poisson point processes $\{\mathcal{D}_j^k\}$ as explained in Section 2. For each finite initial location $\sigma(j,0)$ we construct an auxiliary exclusion process $\zeta^j = (\zeta^j(i,t) : i \in \mathbf{Z}_-, t \geq 0)$. Initially

$$\zeta^j(i,0) = \sigma(j,0) + i \text{ for } i \in \mathbf{Z}_-. \tag{4.1}$$

The particles $\zeta^j(i)$ for $i > 0$ are not needed. According to our convention we may think that they reside permanently at ∞. Thus ζ^j is a process that is a spatial translate of the process started from (3.9).

For each process ζ^j, the dynamical description is the same as for σ: Jumps of size k are attempted independently at rate β_k. Each jump is carried as far as possible without violating the exclusion and no-passing rules. Each process ζ^j uses the same collection of Poisson processes $\{\mathcal{D}_i^k\}$ as does σ, but with a translation of the index:

Jump rule for process ζ^j. At each epoch τ of $\mathcal{D}_{i+j}^k$, set

$$\zeta^j(i,\tau) = \min\left\{\zeta^j(i,\tau-) + k\,,\, \zeta^j(i+1,\tau-) - 1\right\}. \tag{4.2}$$

Between epochs of $\mathcal{D}_{i+j}$, $\zeta^j(i)$ stays constant.

Subject to this rule, the processes $\{\zeta^j\}$ are constructed exactly as σ. The processes σ and $\{\zeta^j\}$ are all defined on the same probability space $(\Omega, \mathcal{F}, P)$ of the Poisson jump times. If this probability space also supports a random initial configuration $\sigma(0)$, then the processes $\{\zeta^j\}$ are functions of both $\sigma(0)$ and $\{\mathcal{D}_i^k\}$, through the translations (4.1) and the jump rule.

This arrangement where several processes are defined on a common probability space to facilitate their direct comparison is known as a *coupling*. It should be evident that the processes are invisible to each other, i.e. particles of one process do not interfere with the jumps of the other processes.

The key property of this coupling is the microscopic counterpart of (3.8):

Lemma 4.1 *The equality*

$$\sigma(i,t) = \inf_{j:j\geq i} \zeta^j(i-j,t) \tag{4.3}$$

holds for all $i \in \mathbf{Z}$ and $t \geq 0$, almost surely.

Proof The exclusion rule (2.1) and (4.1) imply that (4.3) holds at time 0. The jump rules ensure that for each $j \geq i$, particles $\zeta^j(i-j)$ and $\sigma(i)$ attempt k-jumps at common time points, namely at the epochs of $\mathcal{D}_i^k$. Thus for a fixed i the validity of (4.3) can change only at epochs of $\mathcal{D}_i$.

Fix i_1 and t_1. To prove that (4.3) holds for $i = i_1$ up to time t_1, use part (c) of assumption (2.5) to find i_0 such that $-i_0 < i_1 < i_0$, and so that $\mathcal{D}_{-i_0}$ and $\mathcal{D}_{i_0}$ have no epochs in the time interval $[0, t_1]$. By parts (a) and (b) of (2.5), we can do induction on the finitely many epochs in $\cup_{-i_0<i<i_0}\mathcal{D}_i \cap [0, t_1]$. So suppose τ is an epoch of $\mathcal{D}_i^k$ for some $k \geq 1$ and some i between $-i_0$ and i_0, and assume that (4.3) holds for all i between $-i_0$ and i_0 and all $t < \tau$.

First we show that $\leq$ holds in (4.3) right after the jump epoch τ. Let $j \geq i$ be arbitrary. By the induction assumption, $\zeta^j(i-j+1, \tau-) \geq \sigma(i+1, \tau-)$, and consequently $\zeta^j(i-j)$ can jump at least as far as $\sigma(i)$ can. Since before the jump we have $\zeta^j(i-j, \tau-) \geq \sigma(i, \tau-)$, and both particles attempt a k-jump, the inequality $\zeta^j(i-j, \tau) \geq \sigma(i, \tau)$ holds also at time τ after the jump has been completed. The same argument works for all $j \geq i$.

Next we show that right after τ there is some $j \geq i$ that satisfies $\zeta^j(i-j, \tau) = \sigma(i, \tau)$. If $\sigma(i)$ was allowed to jump the full k steps, any j such that $\zeta^j(i-j, \tau-) = \sigma(i, \tau-)$ will do, and by induction there is some such j.

Suppose on the contrary that $\sigma(i)$ was at least partially blocked by $\sigma(i+1)$ at time $\tau-$, so after the jump we have $\sigma(i, \tau) = \sigma(i+1, \tau) - 1$. By induction, we can pick $j \geq i+1$ such that $\zeta^j(i-j+1, \tau-) = \sigma(i+1, \tau-)$. By the earlier part of the proof, $\zeta^j(i-j, \tau) \geq \sigma(i, \tau)$. On the other hand, the exclusion rule for the process ζ^j stipulates that $\zeta^j(i-j, \tau) \leq \zeta^j(i-j+1, \tau) - 1 = \sigma(i+1, \tau) - 1 = \sigma(i, \tau)$. (Notice that by part (b) of assumption (2.5) there is no epoch τ in $\mathcal{D}_{i+1}$.) This shows that after the jump we necessarily have the equality $\zeta^j(i-j, \tau) = \sigma(i, \tau)$. □

By subtracting off the centering at $\sigma(j, 0)$ from ζ^j we obtain a family of processes $\{\xi^j\}$ that are all equal in distribution. Define

$$\xi^j(i, t) = \zeta^j(i, t) - \sigma(j, 0) \text{ for all } i, j, \text{ and } t. \tag{4.4}$$

Then we can turn the coupling equality (4.3) into a statement that is a microscopic version of the Hopf-Lax formula (3.5):

$$\sigma(i, t) = \inf_{j: j \geq i} \left\{\sigma(j, 0) + \xi^j(i-j, t)\right\}. \tag{4.5}$$

The processes $\{\xi^j\}$ are identically distributed on account of the requirement that $(\sigma(j, 0))$ and the Poisson processes $\{\mathcal{D}_j^k\}$ are independent. The advantage of (4.5) over (4.3) is that the effects of $(\sigma(j, 0))$ and $\{\mathcal{D}_j^k\}$ have been separated in two distinct terms inside the braces.

5 The tagged particle limit

As a corollary of the coupling we obtain the limit discussed heuristically in Section 3. The precise assumptions are as follows: There is a sequence of processes σ_n constructed as in Section 2, and an increasing function u_0 on $\mathbf{R}$. At time 0, the convergence

$$\lim_{n \to \infty} n^{-1} \sigma_n([ny], 0) = u_0(y) \tag{5.1}$$

holds in probability for each $y \in \mathbf{R}$. More precisely, if $(\Omega_n, \mathcal{F}_n, P_n)$ is the probability space on which the process σ_n is defined, then

$$\lim_{n \to \infty} P_n\left(|\sigma_n([ny], 0) - nu_0(y)| \geq n\varepsilon\right) = 0 \tag{5.2}$$

for each $y \in \mathbf{R}$ and $\varepsilon > 0$.

Theorem 5.1 *Under assumption* (5.1), *the convergence*

$$\lim_{n\to\infty} n^{-1}\sigma_n([nx], nt) = u(x,t) \tag{5.3}$$

holds in probability for each $x \in \mathbf{R}$ and $t > 0$. The deterministic limit $u(x,t)$ can be macroscopically defined by

$$u(x,t) = \inf_{y:y\geq x}\left\{u_0(y) + tg\left(\frac{x-y}{t}\right)\right\}, \tag{5.4}$$

where g is a certain convex function on $(-\infty, 0]$ determined by the rules of the exclusion process.

The meaning of the limit in (5.3) is the same as above for (5.1), namely that

$$\lim_{n\to\infty} P_n\left(|\sigma_n([nx], nt) - nu(x,t)| \geq n\varepsilon\right) = 0 \tag{5.5}$$

for each $\varepsilon > 0$. By the discussion about Hamilton-Jacobi equations in Section 3, $u(x,t)$ can be equivalently characterized as the unique viscosity solution of $u_t = f(u_x)$ with initial data $u(x,0) = u_0(x)$, where f is the negative of the convex conjugate of g. The theorem says that in this exclusion system, a tagged particle has a well-defined macroscopic velocity $f(u_x)$ as long as initially the particle distribution is sufficiently regular as specified by assumption (5.1).

Proof of Theorem 5.1 To prove Theorem 5.1, we forge a rigorous connection between the matching microscopic and macroscopic descriptions (4.5) and (5.4). For each n, we have the processes ζ_n^j and ξ_n^j, defined as in Section 4 relative to the process σ_n. All the processes ξ_n^j are identical in distribution, so as long as we discuss distributional properties, we may leave out the sub- and superscript n and j. The function g that appears in the theorem is defined by the limit of the next lemma.

Lemma 5.2 *There exists a finite, continuous, convex function g on $(-\infty, 0]$ such that*

$$\lim_{n\to\infty} n^{-1}\xi([nx], nt) = tg(x/t) \tag{5.6}$$

in probability, for each $x \leq 0$ and $t > 0$.

Proof of Lemma 5.2 We start by deriving a useful subadditivity. Assume the process $\xi(i,t)$ has been constructed on a probability space $(\Omega, \mathcal{F}, P)$ as explained in Section 2, so that particle $\xi(i)$ gets its jump commands from the labeled Poisson process $\mathcal{D}_i$. To perform comparisons, we explicitly include the particles that are permanently at infinity: $\xi(i,t) = \infty$ for all $i > 0$ and $t \geq 0$.

Fix an integer $h \in \mathbf{Z}_-$ and a time $s > 0$. Construct two new processes σ' and σ'' on the same probability space $(\Omega, \mathcal{F}, P)$ as follows: The initial configurations are

$$\sigma'(i,0) = \xi(h+i, s) \text{ for all } i \in \mathbf{Z} \tag{5.7}$$

and

$$\sigma''(i,0) = \begin{cases} \infty & \text{if } i > 0, \\ \xi(h,s) + i & \text{if } i \leq 0. \end{cases} \tag{5.8}$$

We stipulate that particles $\sigma'(i)$ and $\sigma''(i)$ read their jump commands from $(\mathcal{D}_{h+i} - s) \cap [0, \infty)$. In other words, σ' and σ'' read the Poisson processes $\{\mathcal{D}_i\}$ after first translating the index by h and time by s. By Lemma 2.1,

$$\sigma'(i,t) \leq \sigma''(i,t) \text{ for all } i \in \mathbf{Z} \text{ and } t \geq 0. \tag{5.9}$$

Since σ' continues ξ with translated index and time, $\sigma'(i,t) = \xi(h+i, s+t)$. For σ'' we can write $\sigma''(i,t) = \xi(h,s) + \widetilde{\xi}(i,t)$ where $\widetilde{\xi}$ is equal in distribution to ξ but independent of it. Thus (5.9) turns into

$$\xi(h+i, s+t) \leq \xi(h,s) + \widetilde{\xi}(i,t)\,, \tag{5.10}$$

where the random variables on the right are independent.

Suppose first that x in (5.6) is a nonpositive integer, $x = i$. Let F_n be the distribution function of $\xi(ni, nt)$. Then (5.10) gives $F_{n+m} \geq F_n * F_m$. By the Kesten-Hammersley lemma from subadditive ergodic theory there exists a function $\gamma(i,t)$ such that

$$\lim_{n\to\infty} n^{-1}\xi(ni, nt) = \gamma(i,t) \tag{5.11}$$

in probability, for all $i \in \mathbf{Z}_-$ and $t \geq 0$. A proof of this lemma can be found on p. 20 in [11]. The assumption of finite second moments required by this lemma is satisfied by assumption (2.3).

The rest of the proof consists of extending the limit to all $x \leq 0$, by using regularity and monotonicity properties of the process and $\gamma(i,t)$. It is obvious that $\gamma(i,t)$ is nondecreasing in both i and t. For $i = 0$ we have $\gamma(0,t) = B_1 t$, because B_1 is the speed of an unobstructed particle. Furthermore

$$i \leq \gamma(i,t) \leq i + B_1 t \tag{5.12}$$

for all $i \in \mathbf{Z}_-$ and $t \geq 0$, because $\xi(i,0) = i$ and B_1 is the maximal average speed.

Next we argue the t-continuity of $\gamma(i,t)$. During the time interval $(nt, nt+n\varepsilon]$ particle $\xi(ni)$ attempts a Poisson$(B_0 n\varepsilon)$ number of jumps. The attempted jump sizes are i.i.d. random variables $\{X_i\}$ with common distribution $P(X_i = k) = \beta_k/B_0$. Due to the exclusion and no-passing rules, the actual jumps are at most X_i, so we can write

$$\xi(ni, nt+n\varepsilon) \leq \xi(ni, nt) + \sum_{i=1}^{N} X_i \tag{5.13}$$

where $N \sim$ Poisson$(B_0 n\varepsilon)$. It follows that

$$\gamma(i, t+\varepsilon) \leq \gamma(i,t) + B_1\varepsilon\,. \tag{5.14}$$

In particular, $\gamma(i,t)$ is Lipschitz continuous in t.

A homogeneity holds: If m is a positive integer, then

$$\begin{aligned} \gamma(mi, mt) &= \lim_{n\to\infty} n^{-1}\xi(nmi, nmt) \\ &= m \cdot \lim_{n\to\infty} (nm)^{-1}\xi(nmi, nmt) \\ &= m\gamma(i,t)\,. \end{aligned}$$

From this it follows that we can define unambiguously for rational $r \leq 0$

$$\gamma(r,t) = \frac{1}{m}\gamma(mr, mt)\,, \tag{5.15}$$

where m is any positive integer such that $mr \in \mathbf{Z}_-$.

We need to prove that the extension (5.15) makes sense for the process, namely that

$$\lim_{n\to\infty} n^{-1}\xi([nr], nt) = \gamma(r,t) \tag{5.16}$$

for rational $r < 0$ and all $t > 0$. If n runs along the subsequence $n = km$, $k = 1, 2, 3, \ldots$, where m is the integer in (5.15), then (5.16) is valid. For other values of n we interpolate using the monotonicity of the process. No-passing and total asymmetry of the dynamics imply that

$$\xi(i,s) \le \xi(j,s) \le \xi(j,t)$$

for all $i \le j$ and $s \le t$. Given n, let $k = k(n)$ be the unique integer that satisfies $km \le n < (k+1)m$. Let $\varepsilon > 0$. Then for large enough n,

$$\begin{aligned} \xi\left((k+1)mr, (k+1)m(t-\varepsilon)\right) &\le \xi([nr], nt) \\ &\le \xi\left(kmr, km(t+\varepsilon)\right), \end{aligned}$$

from which, upon dividing by n and passing to the limit,

$$\begin{aligned} \frac{1}{m}\gamma\left(mr, m(t-\varepsilon)\right) &\le \liminf_{n\to\infty} n^{-1}\xi([nr], nt) \\ &\le \limsup_{n\to\infty} n^{-1}\xi([nr], nt) \\ &\le \frac{1}{m}\gamma\left(mr, m(t+\varepsilon)\right). \end{aligned}$$

These inequalities cannot be taken literally since the limits have to be interpreted in the sense of convergence in probability. But the point should be clear: By the continuity (5.14) and the definition (5.15), the limit (5.16) is valid.

Next one checks that the homogeneity extends to rationals:

$$\gamma(rx, rt) = r\gamma(x,t) \tag{5.17}$$

for rational $x \le 0$, rational $r \ge 0$, and all $t > 0$. From (5.10) follows subadditivity

$$\gamma(x+y, s+t) \le \gamma(x,s) + \gamma(y,t), \tag{5.18}$$

first for integers x, y and then by (5.15) for rational x, y. (5.17) and (5.18) together imply convexity, and also continuity in the x-variable, for rational x.

The final extension of γ is

$$\gamma(x,t) = \sup\{\gamma(r,t) : r \in \mathbf{Q}, r < x\}. \tag{5.19}$$

By the continuity in the x-variable, for rational x this definition gives the old value $\gamma(x,t)$, so it is a sensible extension. Again, one checks that homogeneity (5.17) and subadditivity (5.18) hold, this time for all r, x, y. Convexity follows, and convexity implies continuity in the open quadrant $\{x < 0, t > 0\}$ (Theorem 10.1 in [6]). Continuity up to the boundary requires separate arguments. And again, one checks that the limit (5.16) holds also for irrational r. This follows as it did for rational r, by the monotonicity of the process and the continuity of the limit.

Finally, define $g(x) = \gamma(x, 1)$. By homogeneity $\gamma(x,t) = tg(x/t)$. □

We return to the proof of Theorem 5.1. Fix (x,t). From the coupling (4.5) we get

$$n^{-1}\sigma_n([nx], nt) = \inf_{i:i\ge[nx]} \left\{n^{-1}\sigma_n(i,0) + n^{-1}\xi_n^i\left([nx]-i, nt\right)\right\}. \tag{5.20}$$

By the hypothesis of Theorem 5.1 and by Lemma 5.2, for $i = [ny]$ the random variable inside the braces converges to the quantity inside the braces in (5.4). Consequently we get

$$\lim_{n\to\infty} P_n\left(n^{-1}\sigma_n([nx], nt) \le u(x,t) + \varepsilon\right) = 1 \tag{5.21}$$

for any $\varepsilon > 0$.

For the converse we need some further estimation. The first step is to restrict the variable i in (5.20) to a range of order n. This is achieved by the next lemma. Define

$$w_n(z) = \min_{i:[nx]\le i\le [nz]} \left\{\sigma_n(i,0) + \xi_n^i\left([nx]-i, nt\right)\right\}. \tag{5.22}$$

Lemma 5.3 *For large enough $z > x$, there exists a constant $C = C(z) > 0$ such that*

$$P_n\left(\sigma_n([nx], nt) \ne w_n(z)\right) \le e^{-Cn} \tag{5.23}$$

for all large enough n.

Proof Suppose particle $\xi_n^{[nz]}([nx]-[nz])$ has not moved by time nt, in other words that $\xi_n^{[nz]}([nx]-[nz], nt) = [nx]-[nz]$. Let $i > [nz]$. By the exclusion rule for σ_n and because

$$\xi_n^i([nx]-i, nt) \ge \xi_n^i([nx]-i, 0) = [nx]-i\,,$$

we have

$$\begin{aligned} & \sigma_n(i,0) + \xi_n^i([nx]-i, nt) \\ \ge\ & \sigma_n([nz],0) + i - [nz] + [nx] - i \\ =\ & \sigma_n([nz],0) + [nx] - [nz] \\ =\ & \sigma_n([nz],0) + \xi_n^{[nz]}([nx]-[nz], nt)\,. \end{aligned}$$

The conclusion is that the terms $i > [nz]$ cannot contribute to the infimum in (5.20).

Thus to prove the lemma, it suffices to show that for large enough z, the probability that $\xi_n^{[nz]}([nx]-[nz])$ has moved by time nt is at most e^{-Cn}. This is clear again on account of the exclusion and no-passing rules: The random time when $\xi_n^{[nz]}([nx]-[nz])$ first moves has the distribution of a sum of $[nz]-[nx]+1$ i.i.d. exponential random variables with rate B_0, because after $\xi(j)$ first moves, $\xi(j-1)$ waits an Exp(B_0)-distributed independent time to make its first jump, and these times are added up over $j = 0, -1, -2, \ldots, [nx]-[nz]$. It is sufficient to choose z so that $(z-x)B_0^{-1} > t+\varepsilon$ for some $\varepsilon > 0$. □

Fix z so that (5.23) holds, and let $\varepsilon > 0$. Pick a partition

$$x = y_0 < y_1 < y_2 < \cdots < y_m = z$$

so that

$$\left| tg\left(\frac{x-y_{k+1}}{t}\right) - tg\left(\frac{x-y_k}{t}\right)\right| \le \frac{\varepsilon}{4} \tag{5.24}$$

for $k = 0, 1, \ldots, m-1$. If we could restrict the minimum in (5.22) to the $m+1$ points $i = [ny_k]$, then passing to the limit inside or outside the minimum (now over a fixed finite number of random variables) would make no difference. To control the error of this simplification, we need to control what happens inside the braces in (5.22) for $[ny_k] < i < [ny_{k+1}]$. This can be done with couplings:

Lemma 5.4 *For all n and $i \le j_0 \le j_1$, the inequality $\xi_n^{j_0}(i-j_0,t) \ge \xi_n^{j_1}(i-j_1,t)$ holds almost surely for all times $t \ge 0$.*

Proof Fix n and $j_0 \le j_1$. Consider the processes

$$\sigma'(i,t) = \xi_n^{j_1}(i - j_1, t) \text{ and } \sigma''(i,t) = \xi_n^{j_0}(i - j_0, t),\ i \in \mathbf{Z}.$$

Both $\sigma'(i)$ and $\sigma''(i)$ read their jump commands from $\mathcal{D}_i$. (Recall the general rule from Section 4: particle $\xi^j(i)$ reads jump commands from $\mathcal{D}_{i+j}$.) Initially

$$\sigma'(i,0) = \begin{cases} \infty & \text{if } i > j_1, \\ i - j_1 & \text{if } i \le j_1, \end{cases} \tag{5.25}$$

and

$$\sigma''(i,0) = \begin{cases} \infty & \text{if } i > j_0, \\ i - j_0 & \text{if } i \le j_0. \end{cases} \tag{5.26}$$

By the assumption $j_0 \le j_1$, $\sigma'(i,0) \le \sigma''(i,0)$, and consequently by Lemma 2.1, the inequality $\sigma'(i,t) \le \sigma''(i,t)$ holds for all times $t \ge 0$. □

Apply Lemma 5.4 with $i = [nx]$, $j_0 = i$ between $[ny_k]$ and $[ny_{k+1}]$, and $j_1 = [ny_{k+1}]$, to bound $w_n(z)$ from below by

$$\begin{aligned} w_n(z) &= \min_{0 \le k < m} \min_{i:[ny_k] \le i \le [ny_{k+1}]} \left\{ \sigma_n(i,0) + \xi_n^i\left([nx] - i, nt\right) \right\} \\ &\ge \min_{0 \le k < m} \left\{ \sigma_n\left([ny_k],0\right) + \xi_n^{[ny_{k+1}]}\left([nx] - [ny_{k+1}], nt\right) \right\}. \end{aligned}$$

Multiplied through by n^{-1}, the last line converges in probability by assumption (5.1) and Lemma 5.2. Thus with probability tending to 1 as $n \to \infty$,

$$\begin{aligned} n^{-1} w_n(z) &\ge \min_{0 \le k < m} \left\{ u_0(y_k) + tg\left(\frac{x - y_{k+1}}{t}\right) \right\} - \frac{\varepsilon}{4} \\ &\ge \min_{0 \le k < m} \left\{ u_0(y_k) + tg\left(\frac{x - y_k}{t}\right) \right\} - \frac{\varepsilon}{2} \\ &\ge \inf_{y:y \ge x} \left\{ u_0(y) + tg\left(\frac{x - y}{t}\right) \right\} - \frac{\varepsilon}{2} \\ &= u(x,t) - \frac{\varepsilon}{2}. \end{aligned}$$

This together with Lemma 5.3 gives

$$\lim_{n \to \infty} P_n\left(n^{-1}\sigma([nx], nt) \ge u(x,t) - \varepsilon\right) = 1 \tag{5.27}$$

for any $\varepsilon > 0$. This completes the proof of Theorem 5.1. □

6 Comments and extensions

6.1 Almost sure convergence in Theorem 5.1. It may be possible to upgrade the convergence in Theorem 5.1 to almost sure convergence by developing large deviation estimates for the convergence in Lemma 5.2. Such results can be found in [9] and [10].

6.2 The role of invariant distributions. Theorem 1 does not explicitly identify the function g or the macroscopic velocity f. This can be done when information about steady states is available. Consider the special case of the totally asymmetric simple exclusion process (TASEP) where $\beta_1 = 1$ and $\beta_k = 0$ for $k > 1$. Then it is known that if the interparticle distances are geometrically distributed with mean $v \in [1, \infty)$,

$$P(\sigma(i+1) - \sigma(i) = m) = \left(1 - v^{-1}\right)^{m-1} v^{-1}, \quad m = 1, 2, 3, \dots, \tag{6.1}$$

an individual particle's jumps obey a Poisson process of rate $1 - v^{-1}$. From this information the function g can be calculated with the help of Theorem 5.1, as we next show.

Let each $\sigma_n = \sigma$ be a steady-state process with interparticle distances (6.1), and such that initially $\sigma(0,0) = 0$. Then Theorem 5.1 is valid for $u_0(y) = vy$. Take $x = 0$ and $t = 1$ in (5.4) to get

$$u(0,1) = \inf_{y \geq 0} \{vy + g(-y)\} = -g^*(v), \tag{6.2}$$

where g^* denotes the convex conjugate of g. Note: For the purposes of convex duality, functions should be lower semicontinuous and convex on all of $\mathbf{R}$. (See [6].) The limit of Lemma 5.2 defines g on $(-\infty, 0]$, and we extend g to $\mathbf{R}$ by defining $g(x) = \infty$ for $x > 0$. Then (6.2) is the same as

$$g^*(v) = \sup_x \{xv - g(x)\}. \tag{6.3}$$

Furthermore, for $v < 1$ we obtain $g^*(v) = \infty$ directly from (6.3), by the observation that the slope of g is always at least 1. This is a consequence of the exclusion rule (2.1) and the limit (5.6).

Since $\sigma(0,n)$ is a Poisson$((1 - v^{-1})n)$-variable, we can calculate explicitly that

$$u(0,1) = \lim_{n \to \infty} n^{-1} \sigma(0,n) = 1 - v^{-1}.$$

Thus $g^*(v) = v^{-1} - 1$ for $v \geq 1$. From this we obtain g by performing another convex conjugation: For $x \leq 0$,

$$\begin{aligned} g(x) &= g^{**}(x) \\ &= \sup_v \{xv - g^*(v)\} \\ &= \sup_{v \geq 1} \{xv - v^{-1} + 1\} \\ &= 1 - 2\sqrt{-x}. \end{aligned}$$

The tagged particle velocity is the function $f(v) = 1 - v^{-1}$ obtained above. The relationship of the mean interparticle distance v to the particle density ρ is $\rho = v^{-1}$. As a function of density the velocity is $v(\rho) = f(\rho^{-1}) = 1 - \rho$. From this we get the well-known macroscopic *particle current* of TASEP: $\rho v(\rho) = \rho(1 - \rho)$.

TASEP was the process studied by Rost in 1981 [7] in one of the seminal papers on hydrodynamic limits for asymmetric processes. With an approach completely different from ours he investigated the process ξ and proved a limit theorem related to Lemma 5.2. The use of subadditivity is a common feature of both approaches.

A thorough treatment of TASEP with the method of this paper appears in [8]. In particular, when the invariant distributions are available for explicit calculations, in addition to the law of large numbers it is possible to obtain a (one-sided) explicit large deviation rate function for the tagged particle in a nonequilibrium process.

6.3 More general jump rates. We could admit the following additional feature to the process: Not only is $\sigma(i)$ blocked by the particle in front of him, but he is also held back by the particle behind him, in this particular sense: For some constant $L \geq 1$,

$$\sigma(i) \leq \sigma(i-1) + L \text{ for all } i \in \mathbf{Z}. \tag{6.4}$$

The proof of Theorem 5.1 works with suitable modifications. The initial configurations of the ζ^j-processes have to be changed to $\zeta^j(i) = \sigma(j,0) + i$ for $i \leq 0$, and $\zeta^j(i) = \sigma(j,0) + Li$ for $i > 0$. This feature appears in [9].

Another generalization that this approach can handle is *random rates*. The reader may have noticed that very little depended on how the point processes $\mathcal{D}_i^k$ were actually produced, as long as they had some desirable properties such as those listed in assumption (2.5). In particular, the rates of the processes could vary so that the rate $\beta_{k,i}$ of $\mathcal{D}_i^k$ is a random variable whose realization is fixed. If this randomness is suitably systematic (i.i.d. for example) one would expect that some form of the subadditive ergodicity needed for Lemma 5.2 works again. Examples appear in [9] and [10].

However, so far it has not been possible to admit a more complicated dependence of the rate of jumping on the distance to the nearest particles. By a more complicated dependence we mean a general function $\beta(k,l,r)$ such that jumps of size k are attempted at rate $\beta(k,l,r)$ when there are l and r empty sites to the left and right. In the model of this paper this function is a step function: For a fixed k, $\beta(k,l,r)$ takes only two values depending on whether $k < r$ or $k \geq r$. What appears to fail for more general rates is the proof of Lemma 4.1.

Other shortcomings are that the method has been able to treat only totally asymmetric processes, and processes without passing. A more general asymmetric process would allow jumps both left and right, but with a higher rate for jumping to the right.

References

[1] R. Durrett (1995). Ten lectures on particle systems. Lecture Notes in Mathematics 1608 (Saint-Flour, 1993), 97–201. Springer-Verlag.

[2] L. C. Evans (1998). Partial Differential Equations. American Mathematical Society.

[3] P. A. Ferrari (1996). Limit theorems for tagged particles. Markov Process. Related Fields 2, 17–40.

[4] D. Griffeath (1979). Additive and Cancellative Interacting Particle Systems. Lecture Notes in Mathematics 724, Springer, Heidelberg.

[5] T. M. Liggett (1985). Interacting Particle Systems. Springer-Verlag, New York.

[6] R. T. Rockafellar (1970). Convex Analysis. Princeton University Press.

[7] H. Rost (1981). Non-equilibrium behaviour of a many particle process: Density profile and local equilibrium. Z. Wahrsch. Verw. Gebiete 58, 41–53.

[8] T. Seppäläinen (1998). Coupling the totally asymmetric simple exclusion process with a moving interface. Markov Process. Related Fields 4, 593-628, Proceedings of I Escola Brasileira de Probabilidade, IMPA, 1997.

[9] T. Seppäläinen (1999). Existence of hydrodynamics for the totally asymmetric simple K-exclusion process. Ann. Probab. 27, 361-415.

[10] T. Seppäläinen and J. Krug (1999). Hydrodynamics and platoon formation for a totally asymmetric exclusion model with particlewise disorder. J. Statist. Phys. 95, issue 3/4 529-571.

[11] R. T. Smythe and J. C. Wierman (1978). First-passage percolation on the square lattice. Lecture Notes in Mathematics 671, Springer.

Fields Institute Communications
Volume **27**, 2000

Quantum Mechanics, Linear Boltzmann Equation and Renormalization

Horng-Tzer Yau
Courant Institute of Mathematical Sciences
New York University
New York, NY, 10012, USA
yau@math.nyu.edu

Abstract. We study a quantum particle in a background with lattice structure and impurities modeled by a phonon field and a random potential respectively. In the region of low impurity density (or Boltzmann-Grad) and weak electron-phonon coupling, we show that the phase space density of the electron, defined via the Wigner distribution, converges weakly to a linear Boltzmann equation. The collision kernel is given by the sum of the electron-impurity scattering and the electron-phonon scattering cross sections.

We start with a single electron in a cube of size L in $\boldsymbol{R}^3$ denoted by Λ_L. Let x denote the position of the electron. An electron (we neglect the spin variables) is described by a wave function $\psi(x)$ in the state space $\mathcal{H}_e = L^2(\Lambda_L)$, a Hilbert space with the usual L^2 inner product. The dynamics of the free electron are characterized by the Schrodinger equation

$$i\partial_t\psi_t(x) = H\psi_t(x) \tag{1.1}$$

and the Hamiltonian

$$H = H_e := \frac{-\Delta_x}{2}. \tag{1.2}$$

Here $\psi_t(x)$ is the wave function at time t. The solution of (1.1) defines a unitary group e^{-itH} and it conserves the L^2 norm. This holds for arbitrary Schrodinger equation and is the quantum analogue of the Liouville operator preserving the L^1 norm.

The electron moves in a background described by a phonon field (modeling the lattice vibrations), a random potential modeling obstacles (impurities) and radiation fields (electromagnetic field). We first describe the random potential.

Let $V_0(x)$ denote the potential between an electron and an obstacle. Denote by $\omega_{ob} = (y_\alpha), \alpha = 1, \cdots, N_{ob}$, the configuration of obstacles and let $\rho_b = N_{ob}/L^3$

2000 *Mathematics Subject Classification.* Primary 82C10; Secondary 81C30.
Partially supported by U. S. National Science Foundation grant 9703752.

be the density of the obstacles. The potential due to the obstacles is given by

$$V_{\omega_{ob}}(x) = \sum_{\alpha=1}^{N_{ob}} V_\alpha, \quad V_\alpha(x) := V_0(x - y_\alpha) \tag{1.3}$$

We shall denote $V_{\omega_{ob}}$ by V_{ob} in most cases in this lecture. The evolution of the quantum particle in the random environment generated by these obstacles is given by the Schrodinger equation (1.1) but the Hamiltonian is modified by

$$H = H_e + H_{e-ob}, \quad H_{e-ob} = V_{ob}(x). \tag{1.4}$$

Here V_{ob} is a multiplication operator. This is the quantum analogue of a classical Lorentz gas.

The obstacles are considered randomly placed in the space. Hence the natural law is the Poisson distribution. Since we fix the number of obstacles, it is simply the uniform distribution, namely, there are N_{ob} obstacles with coordinates y_α, $\alpha = 1, \cdots, N_{ob}$ such that each y_α is distributed by the law $L^{-3}dx_\alpha$ in the cube of size L. If we denote the expectation with respect to the obstacles by $\mathbf{E}_{ob}$, then

$$\mathbf{E}_{ob} = \prod_{\alpha=1}^{N_{ob}} \left(L^{-3} \int dy_\alpha \right)$$

Phonon problems

The dynamics of m free particles (phonons) with momentum $k_1, \cdots, k_m$ are governed by the Schrodinger equation

$$i\partial_t \phi_t(k_1, \cdots, k_m) = H_{ph}^m \phi_t(k_1, \cdots, k_m) \tag{1.5}$$

and the Hamiltonian (as a multiplication operator in momentum space)

$$H_{ph}^m(k_1, \cdots, k_m) = \sum_{j=1}^{m} \omega(k_j^2) \tag{1.6}$$

where $\omega(k)$ is the dispersion law and we shall not need its precise form in this lecture. One can take for example $\omega(k) = |k|$. Since we are in a cube of size L, the momentum k_j is in the dual lattice of the cube, i.e., $k_j \in \Lambda_L^{dual} = 2\pi(n_1/L, n_2/L, n_3/L)$ for some integers n_1, n_2, n_3. We shall take the limit $L \to \infty$ and thus k_j eventually can be any vector in $\boldsymbol{R}^3$. The state space of m-phonon wave functions, $\mathcal{H}_{ph}^m$ is the subspace of symmetric functions in $\ell^2[(\Lambda_L^{dual})^m]$. Again, it is a Hilbert space with the usual ℓ^2 inner product.

The combined dynamics of an electron and m phonons is thus given by

$$i\partial_t \phi_t(x; k_1, \cdots, k_m) = \left[H_e \otimes 1 + 1 \otimes H_{ph}^m \right] \phi_t(x; k_1, \cdots, k_m) \tag{1.7}$$

We shall abuse the notation to write H_e for $H_e \otimes 1$ and H_{ph}^m for $1 \otimes H_{ph}^m$. The interactions between the electron and phonons are via emission and absorption of phonons. So the number of phonons is not fixed and it is more convenient to consider the space of phonons with indefinite numbers. The state space of phonons, called the Fock space of phonons, is thus the direct sum of $\mathcal{H}_{ph}^m$:

$$\mathcal{H}_{ph} := \oplus_{m=0}^{\infty} \mathcal{H}_{ph}^m.$$

The free phonon Hamiltonian H_{ph} is the direct sum of Hamiltonian H_{ph}^m with the obvious definition that the restriction of H_{ph} to the m phonon subspace $\mathcal{H}_{ph}^m$ is

H^m_{ph}. The full wave function of the electron and phonons is given by $\Psi = (\Psi^m)_{m=0}^{\infty}$ with

$$\Psi^m = \Psi^m(x; k_1, \cdots, k_m)$$

The state space of the wave function Ψ is $\mathcal{H}_e \otimes \mathcal{H}_{ph}$.

Define the phonon annihilation and creation operators c_k^+ and c_k by

$$(c_k\Psi)^m(x; k_1, \cdots, k_m) = \sqrt{m+1}\Psi^{m+1}(x; k, k_1, \cdots, k_m) \tag{1.8}$$

$$(c_k^+\Psi)^m(x; k_1, \cdots, k_m) = \frac{1}{\sqrt{m}}\sum_{j=1}^{m}\Psi^{m-1}(x; k_1, \cdots, \widehat{k_j}, \cdots, k_m)\delta(k_j - k) \tag{1.9}$$

where we set $(c_k^+\Psi)^{m=0} = 0$ by definition. The interactions between the electron and phonons are very complicated. They are usually modeled by the interaction Hamiltonian

$$H_{e-ph} = i\lambda_{e-ph}\int G(k)\left[\, e^{ikx}c_k - e^{-ikx}c_k^+ \,\right] \tag{1.10}$$

where $G(k)$ models effective electron-phonon interactions and λ_{e-ph} is the electron-phonon coupling constant. The dynamics of an electron coupled to phonons is thus given by

$$i\partial_t\Psi_t(x; k_1, \cdots, k_m, \cdots,) = H\Psi_t(x; k_1, \cdots, k_m, \cdots,) \tag{1.11}$$

where

$$H = H_e + H_{ph} + H_{e-ph} \tag{1.12}$$

is the full Hamiltonian. This is just a system of infinitely coupled equations. If we set $G(k) = |k|^{-1}$ and $\omega(k)$ to be a constant, we have the well-known Frohlich Hamiltonian describing a polaron.

We can put these two settings together to have an electron in a random environment and a phonon background. The equation is still (1.11) but the Hamiltonian is now

$$H = H_e + H_{e-ob} + H_{ph} + H_{e-ph}. \tag{1.13}$$

Here we have assumed that the impurities have no influence on the phonons.

Suppose we are again in the one electron case without phonons. The physics of quantum systems are given by expected value of observables with respect to the wave function:

$$\langle \psi, O\psi \rangle.$$

The observables O can be, say, some multiplication operators, or some differential operators or more generally pseudo-differential operators. Notice that we have to square the wave function and thus introduces some nonlinearity. This is one of the main difference between the classical and quantum dynamics. Though both are linear, the physics of quantum dynamics are governed by the square of the wave function and thus nonlinear in a sense. We can remove this nonlinearity by going into the density matrix formulation. Define for any wave function ψ the associated density matrix by

$$\gamma_\psi = \pi_\psi,$$

the projection operator onto the wave function ψ in $\mathcal{H}_e$. With these notations, we can represent the expected value of the observables by

$$\langle \psi, O\psi \rangle = Tr\ \gamma_\psi O$$

where Tr denotes the trace in $\mathcal{H}_e$. Furthermore, the Schrodinger equation is equivalent to

$$i\partial_t \gamma_t = [\ \gamma_t, H\ ] \tag{1.14}$$

where $\gamma_t = \pi_{\psi_t}$ and $[A, B] = AB - BA$ is the commutator. Notice the similarity between this equation and the classical dynamics where the commutator is replaced by the Poisson bracket. The density matrix formulation is linear in the sense that both the expected value of the observables and the evolution are linear in the density matrix. The price to pay is now we have to solve an operator equation instead of the usual partial differential equation. The two formulations, however, encounter similar technical difficulties in most cases. The density matrix formulation is necessary to study quantum systems at positive temperatures when the wave function description is no longer valid. The generalization of this setup to include phonons is straightforward and we shall not repeat it.

Our goal is to study the average dynamics of the electron for a typical random environment of obstacles when the phonons are in equilibrium. The equilibrium states of the phonons at the time $t = 0$ are simply the Gibbs states

$$\gamma_{ph,0} = Z^{-1} \exp[\ -\beta H_{ph}\ ] \tag{1.15}$$

where Z is the normalization so that $\gamma_{ph,0}$ is a probability measure. The Gibbs measure $\gamma_{ph,0}$ is characterized by the following property: The events of finding n phonons with momentum k are independent and the probabilities are given by

$$Z^{-1} e^{-\beta\omega(k)n}$$

The expected number of particles with momentum k can be computed explicitly as

$$E^{\gamma_{ph,0}}[\ N_k\ ] = \frac{e^{-\beta\omega(k)}}{[1 - e^{-\beta\omega(k)}]} := n_{ph}(k; \beta,) \tag{1.16}$$

We shall fix β and this index will be omitted.

Suppose the initial state of the electron-phonon system is the tensor product of the electron density matrix $\gamma_{e,0}$ with the phonon density matrix:

$$\gamma_0 = \gamma_{e,0} \otimes \gamma_{ph,0}$$

The electron density matrix can be just

$$\gamma_{e,0} = \pi_{\psi_0}$$

for some initial wave function ψ_0 of the electron. We use the Duhamel formula to write the evolution of the electron-phonon system (1.11) as

$$e^{-itH} = e^{-it(H_e + H_{ph})} - i \int_0^t e^{-i(t-s)(H_e + H_{ph})} H_{e-ph} e^{-is(H_e + H_{ph})} ds + \cdots$$

We now take the partial trace over the phonons, namely, we integrate out the phonons. This can be performed easily since both the phonon Hamiltonian H_{ph} (1.6) and the electron-phonon interaction H_{e-ph} (1.10) are diagonalized in the momentum space representation. The resulting effective (marginal) evolution of the electron in the electron-phonon system is thus given by

$$i\partial_t \psi_t = H\ \psi_t, \quad H = H_e + \lambda_{e-ph} V_{\omega_{ph}} \tag{1.17}$$

where the potential $V_{\omega_{ph}}(x, t)$ is a centered Gaussian field. The subscript ω_{ph} denotes the random parameter of the Gaussian field and has no active meaning.

Hence in most cases we shall denote $V_{\omega_{ph}}$ by V_{ph}. The covariance of V_{ph} can be computed explicitly from the dynamics and the initial state of the phonons

$$\mathbf{E}_{ph} \hat{V}_{ph}(p,t)\overline{\hat{V}_{ph}(q,s)} = \delta(p-q)\left[\; R_1^2(p)e^{-i(t-s)\omega(p)} + R_2^2(p)e^{i(t-s)\omega(p)} \;\right] \quad (1.18)$$

where ω is the phonon dispersion law (1.6) and

$$R_1^2(p) = |\hat{G}(p)|^2\,[\;1+n_{ph}(p)\;]\,, \qquad R_2^2(p) = |\hat{G}(p)|^2 n_{ph}(p) \quad (1.19)$$

with $n_{ph}(p)$ the expected number of phonons (1.16) given by the initial state (1.15). The function $\hat{G}$ is the Fourier transform of G appeared in the Hamiltonian H_{e-ph} (1.10). Notice that we use ω to denote both random parameter and the dispersion law of phonons.

Thus from now on we can assume that we have a one particle problem with

$$i\partial_t \psi_t = H\;\psi_t, \quad H = H_e + H_{e-ob} + \lambda_{e-ph}V_{ph} \quad (1.20)$$

with H_{e-ob} described by (1.4) and V_{ph} by (1.18). Recall that H_{e-ob} depends on the location of the obstacles ω_{ob}. Denote by $\mathbf{E}_{ph}$ the expectation w.r.t. the Gaussian law and by $\mathbf{E}_{ob}$ with respect to the law of the obstacles. The quantity we are interested in, the average dynamics of the electron, is characterized by

$$\mathbf{E}\;\gamma_t := \mathbf{E}_{ob}\mathbf{E}_{ph}\gamma_t$$

where γ_t is the density matrix associated with the wave function ψ_t. We shall denote by ω the collection of the random parameter s ω_{ph} and ω_{ob}. The electron wave function ψ thus depends on ω and will be denoted by ψ_ω. We can ask more refined question with no expectations, but even with the expectation most basic questions are still open.

We can write the density matrix as a kernel $\gamma(x,y)$ such that

$$(\gamma g)(x) = \int \gamma(x,y)g(y)dy$$

The (inverse) Fourier transform of $\gamma(x,y)$ in its difference variable, $y-x$ is called the Wigner distribution. More precisely,

$$W_\gamma(x,v) = \int e^{ivz}\;\gamma(x-\frac{z}{2}, x+\frac{z}{2})\;dz$$

If γ is the projection operator of ψ then

$$\gamma_\psi(x,y) = \psi(x)\overline{\psi}(y)$$

and

$$W_\psi(x,v) = \int e^{ivz}\;\psi\Big(x-\frac{z}{2}\Big)\overline{\psi\Big(x+\frac{z}{2}\Big)}\;dz$$

The Wigner distribution $W_\gamma(x,v)$ represents the 'probability' of finding an electron at position x with velocity v. We have put the probability in quotation mark because typically the Wigner distribution has no definite sign and thus cannot be a probability density. But its proper mollification does. We shall prove that the weak limit of the rescaled Wigner function is nonnegative and thus can be considered as the phase space density.

Let ε be the atomic scale parameter. Define the macroscopic coordinates (X,T) by

$$(X,T) := (x\varepsilon, t\varepsilon)$$

Note that $v = V$, namely, the velocity is not rescaled. In physical reality $\varepsilon \sim 10^{-6}$, and in this model we take the idealized $\varepsilon \to 0$ limit. Our goal is to study the time evolution of the rescaled Wigner distribution:

$$W_\gamma^\varepsilon(X,V) := \varepsilon^{-3} W_\gamma\Big(\frac{X}{\varepsilon}, V\Big) \tag{1.21}$$

in the limit $\varepsilon \to 0$. We have chosen the time scale $t \sim \varepsilon^{-1}$. So far we have not fixed the density of the obstacles or the strength of the interaction between electrons and phonons. The first region that we have nontrivial evolution of the electron is

$$\rho_{ob} = N_{ob}/L^3 = \varepsilon, \qquad \lambda_{e-ph} = \sqrt{\varepsilon} \tag{1.22}$$

i.e., the density of the obstacles is of order ε and the coupling constant is of order $\sqrt{\varepsilon}$. The choice that the density of obstacles is proportional to ε is to keep the typical number of scatterings between the electron and the obstacles, $\rho_{ob}\, t$, finite in the limit. This is similar to the Grad limit for the Boltzmann equation and we shall still call it Grad limit. The choice of the coupling constant is harder to explain. Roughly speaking, it is because the effective scattering cross section of a potential λV is actually proportional to λ^2, a standard fact from scattering theory of quantum mechanics. It will be called the weak coupling limit. Our goal is to show that, in this region as $\varepsilon \to 0$, the expected value of the Wigner distribution of the electron satisfies a Boltzmann equation with the collision kernel described in the following.

Recall the linear Boltzmann equation with collision kernel $\sigma(U,V)$ is given by

$$\begin{aligned} &\partial_T F_T(X,V) + V\cdot\nabla_X F_T(X,V) \\ &\quad = \int \sigma(U,V)F_T(X,U)dU - \sigma F_T(X,V)\,, \end{aligned} \tag{1.23}$$

where $\sigma(U,V)$ is the differential scattering cross section and $\sigma := \int \sigma(V,U)dU$ the total cross section. The differential scattering cross section has two contributions, one from the obstacles and one from the field. The one due to the field is

$$\begin{aligned} \sigma_{ph}(U,V) &= 4\pi R_1^2(U-V)\delta\left[\, U^2/2 - V^2/2 - \omega(U-V)\,\right] \\ &\quad + 4\pi R_2^2(U-V)\delta\left[\, U^2/2 - V^2/2 + \omega(U-V)\,\right] \end{aligned} \tag{1.24}$$

where ω is the dispersion law of the phonons (1.6) and R_1, R_2 are defined in (1.19). The first term, $\delta\left[\, U^2/2 - V^2/2 - \omega(U-V)\,\right]$, describes the scattering that the incoming particle is the electron with momentum U and the outgoing particles are the electron with momentum V and the phonon with momentum $U-V$. The momentum transfer between the electrons, $U-V$, becomes the phonon momentum due to the conservation of momentum. The energy conservation then forces the delta function $\delta\left[\, U^2/2 - V^2/2 - \omega(U-V)\,\right]$. In the second term the incoming particles are the electron and the phonon and the outgoing particle is the electron. The interpretation is similar. The differential scattering cross section from the obstacles is described as follows.

We assume that the potential V_0 between the electron and one obstacle is small enough so that the one body Hamiltonian $H_1 := -\Delta/2 + V_0$ has no bound states and both the incoming and outgoing Hilbert spaces are the full space $L^2(\boldsymbol{R}^3)$. Recall the wave operators $\Omega_\mp = \lim_{s\to\infty} e^{\pm isH_0}e^{\mp isH_1}$, where $H_0 := -\Delta/2$. Under certain assumptions on the potential V_0 the kernel of the scattering operator $S = \Omega_-^*\Omega_+$

in the Fourier space can be written as

$$S(u,v) = \delta(u-v) - 2\pi i \delta(u^2 - v^2) T(u,v). \tag{1.25}$$

The first term, $\delta(u-v)$, represents the event that the particle passes through without scattering; the second term defines the differential scattering cross section

$$\sigma_{ob}(u,v) = 4\pi\, \delta(u^2 - v^2) |T(u,v)|^2. \tag{1.26}$$

The delta function $\delta(u^2 - v^2)$ represents the energy conservation of the scattering process and is called the on-shell condition. The total differential scattering cross section is the sum

$$\sigma = \sigma_{ob} + \sigma_{ph}. \tag{1.27}$$

To get some idea of the quantum dynamics, we first look into the classical dynamics. The evolution of the phase space density of a classical Lorentz gas was studied a long time ago and was proved to converge to a linear Boltzmann equation by G. Gallavotti [6] and later on by H. Spohn [9] and Boldrighini, Bunimovich and Sinai [1] with different methods and in stronger senses in probability. The picture changes in the weak coupling limit; the classical dynamics actually converges to a Brownian motion in this case (Kesten, Papanicolaou [8] for $d \geq 3$ and Durr, Goldstein, Lebowitz [2] for $d = 2$ in appropriate scaling). The interesting point is, in the quantum case, the dynamics still converge to a Boltzmann equation. This was first proved by H. Spohn [10] two decades ago when the macroscopic time is small. Spohn's result was improved by Ho, Landau, Wilkins [7] but the essential assumptions stayed the same. The reason that the quantum dynamics gives a different result from the classical dynamics is roughly speaking due to the following observation: In classical dynamics, a weak obstacle will deflect the electron slightly. Thus the total effects after many collisions is the sum of these small deflections. Since the obstacles are randomly distributed, these small deflections can be considered as inde pendent random variables assuming that the dynamics generate no correlations. Thus their sum converges to a Brownian motion. In quantum dynamics, when the strength of the obstacle is small, it is the *probability* that the electron gets scattered becomes small. Once it is scattered, the typical angle is large. Thus it gives the Boltzmann equation. The quantum Lorentz gas in the Grad limit, though appeared to be a complete analogue to the classical case, is actually much more difficult to analyze mathematically than the weak coupling limit. The reason is mainly due to the renormalization needed to obtain the quantum differential scattering kernel appeared in the final Boltzmann equation This means that the Grad limit is not a semi-classical limit in the quantum case and we have obtained a particle picture of quantum mechanics through a non semiclassical limit. We now describe the result of a joint work with L. Erdos [3,4,5]. Our results cover both limits and are global in time.

We shall choose the initial data as

$$\psi_0^\varepsilon(x) := \varepsilon^{-3/2} h(\varepsilon x) \exp\left(\frac{iS(x\varepsilon)}{\varepsilon}\right). \tag{1.28}$$

and $h(X), S(Y)$ are in the Schwarz class such that S has only isolated nondegenerate critical points. It is easy to check that the rescaled Wigner distribution (1.21) converges to

$$W^\varepsilon_{\psi_0^\varepsilon}(X,V) \to |h(X)|^2 \delta(V - \nabla S(X)) := F_0(X,V) \tag{1.29}$$

weakly on $\boldsymbol{R}^6$ as $\varepsilon \to 0$ and thus $F_0(X,V)$ is the initial data for the Boltzmann equation.

Theorem 1.1 *Let* $\rho = \varepsilon$, $\lambda_{e-ph} = \sqrt{\varepsilon}$ *and let* $\psi^{\varepsilon}_{\omega,t}$ *solve the Schrödinger equation (1.20) with initial condition (1.28). Then for any* $T > 0$ *the rescaled Wigner distribution*

$$\mathbf{E} W^{\varepsilon}_{\psi^{\varepsilon}_{\omega,T/\varepsilon}}(X,V) \to F_T(X,V)$$

weakly as $\varepsilon \to 0$ *and* F_T *satisfies the linear Boltzmann equation (1.23) with initial condition* $F_0(X,V)$ *and collision kernel (1.27).*

Sketch of the Proof

We shall work in microscopic coordinates and the index ε will be omitted. We consider only the Grad limit and take $\lambda_{e-ph} = 0$. Our proof is based on the Duhamel formula. We expand e^{-itH} up to the n-th order term and choose n large enough (depending on ε) so that the error term Ω_n becomes negligible in the limit as $\varepsilon \to 0$. The main steps are given in the following.

1. Quantum scattering kernel and renormalization

To get some feeling, we first compare the classical dynamics and the quantum dynamics. Suppose that the particle in a Lorentz gas has one collision with some obstacle. Classically we simply choose a scatterer and the particle collides with it. In quantum mechanics, we have from the Duhamel formula

$$\begin{aligned}
\psi_t &= e^{-itH}\psi_0 = e^{-itH_0}\psi_0 + \psi^1(t) + \Omega_2, \\
\psi^1(t) &= -i\int_0^t e^{-i(t-s)H_0}\, V_\omega\, e^{-isH_0}\psi_0\, ds \\
\Omega_2 &:= \int_{0\le s_j\le t;\sum s_j = t} e^{-is_0 H}\, V_\omega\, e^{-is_1 H_0}\, V_\omega e^{-is_2 H_0}\, \psi_0\, ds_0\, ds_1\, ds_2
\end{aligned}$$

where $V_\omega = V_{\omega_{ob}}$ is the potential given in (1.3) and $H_0 = -\Delta/2$ (We replace the notation of H_e by H_0 because it is traditionally used in the Duhamel formula). The one collision term is the term denoted by $\psi^1(t)$. By definition, the potential can be written as a sum over obstacles,

$$V_\omega = \sum_{\alpha=1}^{N_{ob}} V_\alpha(x)$$

where $V_\alpha(x) = V_0(x - y_\alpha)$. Hence $\psi^1(t)$ can be written as a summation over obstacles as

$$\psi^1(t) := \sum_{\alpha=1}^{N_{ob}} \psi_\alpha(t)$$

where

$$\psi_\alpha(t) = -i\int_0^t ds\, e^{-i(t-s)H_0}\, V_\alpha\, e^{-isH_0}\, \psi_0.$$

Notice that instead of a collision with a chosen scatterer like in classical dynamics, it is now a sum of collisions with all scatterers! Recall that the physical quantities are defined in terms of the square of the wave function. So we first try to get some idea of the size of

$$\mathbf{E}\|\psi_1(t)\|^2 = \sum_{\alpha=1}^{N_{ob}}\sum_{\beta=1}^{N_{ob}} \mathbf{E}\,\langle\, \psi^1_\alpha,\, \psi^1_\beta\,\rangle$$

where the last expectation is with respect to $L^{-3}\int_{\Lambda_L} dy_\alpha L^{-3}\int_{\Lambda_L} dy_\beta$, with a slight abuse of notation. We shall call the term $\alpha=\beta$ the direct term and $\alpha\neq\beta$ the indirect term.

Since the only dependence on the y_α and y_β is from the potential V_α and V_β,

$$\mathbf{E}\langle\psi_\alpha,\psi_\beta\rangle \sim \mathbf{E}\langle V_\alpha, V_\beta\rangle = L^{-6}\int_{\Lambda_L}\int_{\Lambda_L} dy_\alpha dy_\beta V_0(x-y_\alpha)V_0(x-y_\beta) \sim L^{-6}\neq 0$$

even $\alpha\neq\beta$. Though they are small, the number of the off-diagonal terms is much larger than that of diagonal terms. So we would like to have them equal to zero unless $\alpha\neq\beta$. To achieve this, we simply replace the potential V_0 by V_0-c for some constant so that

$$\int_{\Lambda_L}(V_0(x)-c)dx = 0, \quad c = L^{-3}\int_{\Lambda_L} V_0(x)dx \sim L^{-3}.$$

Clearly we then have

$$\mathbf{E}\left\langle V_\alpha,\ V_\beta\right\rangle = 0 \qquad \text{if } \alpha\neq\beta\,.$$

This will force us to change H_0 to H_0+cN_{ob} (N_{ob} is the number of the obstacles) so as not to change the full Hamiltonian. After this procedure, we will have the indirect terms equal to zero. The direct terms can be computed explicitly. Define

$$K(t;p_0,p_1) := -i\int_0^t ds\; e^{-isp_0^2/2}\, e^{-i(t-s)p_1^2/2}\,.$$

Then rather simple computation using the Fourier transform gives

$$\mathbf{E}\|\psi^1(t)\|^2 = \rho_{ob}\int\int dp_0\, dp_1\, |K(t;p_0,p_1)|^2|\widehat{V}(p_0-p_1)|^2|\widehat{\psi}_0(p_1)|^2$$

where ρ_{ob} is the density of the obstacle (1.22) and the hat denotes the Fourier transform. Notice that there is a time integration in K, so K^2 appears to be of order t^2. A more careful computation for the integration in p however shows that it is effectively of order t and we thus have

$$\mathbf{E}\|\psi^1(t)\|^2 \sim \rho_{ob}\, t.$$

Therefore, we have shown that the one collision terms are at least of order similar to the classical dynamics in case $\rho_{ob}\, t$ is small.

The next question is where we get the collision kernel. This in the field theory language is from the resummation of loop diagrams for the propagator. Roughly speaking, we have denoted by ψ_α the collision with the obstacle α. But the physical collision to this obstacle is not just this term. Suppose we only have one obstacle α. We can also expand the wave function by the Duhamel formula to have

$$\begin{aligned} & e^{-it(H_0+V_\alpha)}\psi_0 \\ &= e^{-itH_0}\psi_0 - i\int_0^t ds e^{-i(t-s)H_0}\, V_\alpha\, e^{-isH_0}\,\psi_0 + \cdots \\ &\quad + (-i)^n\int_{0\le s_j\le t;\sum s_j = t} e^{-is_0H_0}\, V_\alpha\cdots V_\alpha\, e^{-is_nH_0}\,\psi_0\, ds_0 ds_1\cdots ds_n + \cdots \end{aligned}$$

The physical scattering with one obstacle is the infinite sum of all these terms in the Duhamel formula. Suppose we can also do this in the many obstacles context and we denote

$$\psi_\alpha^{ren}$$

the true physical collision with the obstacle α. This step is by no means simple as it assumes that the interferences due to other particles are negligible. Assuming this, we can now write the wave function as

$$\psi^{ren,1}(t) := \sum_{\alpha=1}^{N_{ob}} \psi_\alpha^{ren}(t)$$

We want to have as before that

$$\mathbf{E} \left\langle \psi_\alpha^{ren}, \psi_\beta^{ren} \right\rangle = 0 \tag{1.30}$$

for $\alpha \neq \beta$. Now it becomes clear that a subtraction of a constant from V_0 is in fact meaningless. In order to assure that the indirect terms vanish for one collision processes, we have to subtract, roughly speaking, $L^{-3}T(p,p)$ from V. Here T is defined through the wave operator S in (1.25). Notice the subtraction is momentum dependent. This subtraction has to be rebalanced by adding it to the free Hamiltonian H_0. This changes the dispersion law of the free evolution and is the price to pay in order to have (1.30).

2. Permutation incoherence and combinatorics

We can generalize the previous concept of collision with a obstacle to collision with a sequence of obstacles

$$A := (\alpha_1, \alpha_2, \cdots, \alpha_n), \quad \alpha_j \in \omega \text{ for all } j, \quad \alpha_j \neq \alpha_{j+1}.$$

We denote the term associated with this collision sequence by ψ_A^{ren} and we hope that up to small error

$$\psi \sim \sum_A \psi_A^{ren}. \tag{1.31}$$

We again try to bound the norm of the right hand side. Since we have performed the renormalization, we have

$$\mathbf{E} \left\langle \psi_A^{ren}, \psi_B^{ren} \right\rangle = 0$$

unless $\{A\} = \{B\}$ as *sets*. Hence we can assume here that B is a permutation of A. The case $A = B$, as sequences, is called the main terms; the rest are the cross terms. We have to show that the cross terms are of lower order due to phase incoherence and only the main terms contribute. The number of permutation of n obstacles is $n!$. This factor needs to be removed if we wish to use the right hand side of (1.31). Notice that the $n!$ again comes from squaring the wave function.

There is no easy way to remove this factor. One way to overcome this is to use the $1/n!$ coming from the time integrations. Notice that n collisions corresponds to n time integrations. The time integration is ordered, namely, $t_1 < t_2 < \cdots < t_n$. Therefore, there is a factor $1/n!$ from the time integration. We can use this factor to cancel the $n!$ from the combinatorics of the permutations. On the other hand, since $1/n!$ is needed for convergence globally in time, this approach will be restricted to the convergence in short time [10,7]. In fact, apart from the short time restriction this idea has other problems with renormalization as well and it can be carried out only in the weak coupling limit with Gaussian laws.

Our approach [3,4,5] is based on a scheme combining partial integration in time and a cutoff estimate of the Duhamel formula. The cutoff uses the unitarity of the evolution operator e^{-itH} and the triangle inequality. For any function h_s we have

$$\left\| \int_0^t e^{-i(t-s)H} h_s ds \right\|^2 \leq t \int_0^t \left\| e^{-i(t-s)H} h_s \right\|^2 ds \leq t \int_0^t \|h_s\|^2 ds .$$

Notice that the extra factor t which is a price we pay from using the unitarity. Thus once we know an error is small enough by a factor t^{-1}, it can be removed by unitarity of the Schrodinger operator. This allows us to perform some cutoff. A straightforward implementation of this idea will still have difficulties in the combinatoric s, but it is too technical to describe it here. This is the place that we have to use a partial time integration scheme [3,4,5].

3. Computation of Wigner distribution of the main terms

Finally we have to compute the Wigner distribution of the main terms. The computation is somewhat tricky as we have twice as many time integrations in the Wigner distribution (which involves the square of the wave function) as in the classical case. We have to show that half of these time integrations produce the quantum scattering kernel and the other half corresponds to the time integration of classical dynamics. This concludes the sketch of the proof.

References

[1] C. Boldrighini, L. Bunimovich and Y. Sinai: On the Boltzmann equation for the Lorentz gas, *J. Stat. Phys.*, **32**, 477-501, (1983).

[2] Durr, S. Goldstein and J. Lebowitz: Asymptotic motion of a classical particle in a random potential in two dimensions: Landau model. *Commun. Math. Phys.*, **113**, 209-230 (1987).

[3] L. Erdős and H. T. Yau: Linear Boltzmann Equation as Scaling Limit of Quantum Lorentz Gas, *GT-UAB Conf. Proceedings, Contemporary Math.*, **217**, 137-157, 1997

[4] L. Erdős and H. T. Yau: Linear Boltzmann Equation as weak coupling limit of a quantum particle in random field, to appear in *Commun. Pure. Appl. Math.*

[5] L. Erdős and H. T. Yau: Linear Boltzmann Equation as the weak coupling limit of an electron-phonon system, in preparation.

[6] G. Gallavotti: Rigorous theory of the Boltzmann equation in the Lorentz gas, Nota inteerna n. 358, Univ. di Roma (1970).

[7] T. G. Ho, L. J. Landau and A. J. Wilkins: On the weak coupling limit for a Fermi gas in a random potential, *Rev. Math. Phys.*, **5**, 209-298 (1992).

[8] H. Kesten, G. Papanicolaou: A limit theorem for stochastic acceleration, *Commun. Math. Phys.*, **78**, 19-63 (1980).

[9] H. Spohn: The Lorentz process converges to a random flight process, *Commun. Math. Phys.*, **60**, 277-290 (1978).

[10] H. Spohn: Derivation of the transport equation for electorns moving through random impurities, *J. Stat. Phys.*, **17**, p385, (1977).